D1117336

THE MAKING OF STATISTICIANS

The Making of Statisticians

Edited by J. Gani

With 18 Illustrations

Springer-Verlag
New York Heidelberg Berlin

J. Gani
Department of Statistics
University of Kentucky
Lexington, Kentucky, 40506
U.S.A.

Library of Congress Cataloging in Publication Data
The Making of Statisticians.
Includes index.
1. Statisticians—Autobiography. I. Gani, J.M.
(Joseph Mark)
QA276.156.M34 519.5'092'2[B]82-3347

Printed in the United States of America.

9 8 7 6 5 4 3 2 1

ISBN 0-387-90684-3 Springer-Verlag New York Heidelberg Berlin
ISBN 3-540-90684-3 Springer-Verlag Berlin Heidelberg New York

Preface

Like many other scientists, I have long been interested in history. I enjoy reading about the minutiae of its daily unfolding: the coinage, food, clothes, games, literature and habits which characterize a people. I am carried away by the broad sweep of its major events: the wars, famines, migrations, reforms, political swings and scientific advances which shape a society. I know that historians value autobiographical accounts as part of the basic material from which the stuff of history is distilled; this should apply no less to statistical than to political or social history.

Modern statistics is a relatively young science; it was while pondering this fact sometime in 1980 that I realized that many of the pioneers of our field could still be called upon to tell their stories. If, however, biographical material about these eminent statisticians was not gathered, then one might lose the chance to gain insight into the origins of many an important statistical development. The remarkable experience of these colleagues could not be readily duplicated.

Fired by these thoughts, I took it upon myself to plan the framework of this book. In it, eminent statisticians (probabilists are included under this title) would be invited to sketch their lives, explain how they had become interested in probability and statistics, give an account of their major contributions, and possibly hazard some predictions about the future of the subject.

In an attempt to cover the field as broadly as possible, I wrote to more than twenty colleagues outlining my plans; I was fortunate in obtaining positive responses from sixteen of them. Among those who declined were several who had already published their reminiscences, and thought it unwise to do so again. Each of the contributors to this book has made his mark on statistics; each has an original approach to the subject. I was surprised to note how different these individual approaches were and how broad were the backgrounds and interests of the writers; many had worked as actuaries, some had been employed by government, others by industry, all but one had at some time been academics. Yet, in the final analysis, the paths which had led them to statistics appeared to be almost random.

The uniqueness of these eminent statisticians is such that they defy exact classification. There are, however, connecting threads in their work and similarities in their interests which have permitted me to group them into sections reflecting their dominant orientations. These are as follows: 1) probabilists: B. de Finetti, E. Lukacs, and G. Mihoc; 2) statisticians in stochastic processes and independence: M. S. Bartlett and M. Kac; 3) mathematical statisticians: Z. W. Birnbaum, R. C. Bose, W. Hoeffding, and E. J. G. Pitman; 4) statisticians in design and computing: R. L. Anderson, D. J. Finney, and T. Kitagawa: 5) statisticians in industry and econometrics: L. H. C. Tippett and H. Wold; 6) statisticians in demography and medicine: B. Benjamin and H. O. Lancaster.

My classifications are inevitably lacking in precision, and are perhaps dictated by somewhat personal views; I shall take no offense if the authors or readers disagree with them. They are certainly not intended to isolate the contributors into rigid categories; each of them has contributed to several areas of statistics, and could just as easily be placed in a different grouping.

Let me conclude by remarking that, for my own part, I have been fascinated by the wealth of experience, the liveliness, the independence of thought and the humanity which emerge from these autobiographical accounts. I had initially thought to write a lengthier preface in which I would survey major developments and changes on the statistical front over the past 50 years. For example, there is evidence to indicate that the transformation of our educational and university systems from their past relatively élitist form to their present more democratic structure, has had its effect on statisticians. After reading the authors' contributions, however, I felt an overwhelming need for humility. Readers will recognize that these accounts tell the story with far greater authenticity than any survey of mine could have achieved. I am greatly indebted to every one of the sixteen contributors for sharing their lives and their thoughts with me and with the readers of this book. I have learned from them a great deal which I value.

My thanks are also due to Mrs. Kathleen Lyle and Miss Mavis Hitchcock of the Applied Probability Trust Office for their editorial assistance, and to Springer-Verlag for publishing the book on behalf of the Trust.

Lexington, Kentucky
17 September 1981

J. GANI

Contents

1
PROBABILISTS

Bruno de Finetti

Bruno de Finetti was born on 13 June 1906 in Innsbruck, Austria of irredentist Italian parents. One of his grandmothers was Anna Radaelli, niece of General Carlo Radaelli, defender of Venice against the Austrian aggression in 1848–9.

Professor de Finetti attended middle and high schools at Trento, and after graduating there, entered the Polytechnic at Milan. During his third year of study he recognized his preference for a scientific rather than professional training, and joined the Faculty of Mathematics at the newly opened University of Milan. He soon became interested in a problem of mathematical genetics and wrote a paper on the statistical consequences of Mendel's laws under the hypothesis of random mating.

As a result of this paper, his interest in statistics grew, and on completing his degree, he joined ISTAT (the newly formed Central Statistical Bureau) where he was appointed head of the Mathematical Office. He later joined an insurance company, Assicurazioni Generali, in Trieste as an actuary and was involved in the automation of their systems by IBM machines.

He was soon appointed lecturer in the Universities of Trieste and Padua, and later became a full Professor first in Trieste and later in Rome. He has been a very influential figure not only in Italian statistics but also on the international scene. He has written many papers and several books, chief among them *Probability, Induction and Statistics* (1972) and *Theory of Probability* (1970), which has been translated into English (1974–5) and German (1981). Among the many honours conferred upon him was his election as Corresponding and later Full Member of the Accademia dei Lincei. His international stature has been recognized by the recent Meeting on Exchangeability held in 1981 at the Accademia in Rome to celebrate his 75th birthday.

He is married to Renata Errico, and their daughter Fulvia works at IBM Italia, Rome.

Probability and My Life

Bruno de Finetti

1. My Early Life and Education

I was born on 13 June 1906 in Innsbruck, where my father, an engineer, was directing the construction of the alpine electric railroad to Fulpmes (the Stubaithalbahn), planned by him for the Riehl construction company. My parents and grandparents and their families—including myself—were Austrian subjects; but we were also Italian and therefore "irredentists," until the collapse of the Austro-Hungarian Empire in 1918. A particularly strong and enthusiastic irredentist was my paternal grandmother Anna Radaelli; she was the niece of General Radaelli, commander of the defence of Venice against Austria in 1848–49.

I attended primary school and high school in Trento. (This was my mother's home; my father died suddenly in 1912, and she had returned to her native city.) After high school I entered Milan Polytechnic; but while I was in my third year a university with a Faculty of Mathematics was opened in Milan. After listening to some lectures there I realised that I preferred the more theoretical curriculum. So I joined the university, and gained one year, completing my degree in four years instead of the five required at the polytechnic.

This was important for myself and my family, because the depreciation of the Austrian currency, the conversion to Italian lire and the rapid inflation at that time were terrible problems for a family without income.

2. How I Happened to Meet Probability

It is strange that I entered the field of probability through a short article in a popular magazine about Mendel's laws. The author, the biologist Carlo Foà, was in Milan when I was there as a student at the Polytechnic. I wrote a paper on the topic, and sent my manuscript to him. It was about the statistical consequences of Mendel's laws on the diffusion of genetic characteristics under panmixia. He read it and passed it on to his colleagues in mathematics and statistics, and finally to Corrado Gini, president of the newly created Istituto Centrale di Statistica (National Census Bureau). Gini found the paper interesting, and not only published it in his revue *Metron*, but offered me a position in his institute after I graduated. I accepted, and was appointed head of the Mathematical Office.

Later I accepted a position as actuary in the Assicurazioni Generali insurance company in Trieste, where I was chiefly concerned with the automation of all actuarial, statistical and bookkeeping operations and calculations. First Hollerith and later IBM equipment was used.

I was also a Lecturer in Mathematics, Actuarial Mathematics and Probability at the Universities of Trieste and Padova, and later full professor in Trieste and finally in Rome. During this period, I was also a visiting professor at the University of Chicago for a quarter, having been invited by L. Jimmie Savage. He was an incomparable colleague and friend who unfortunately died too soon; it happened after he returned home from a congress in Bucharest where we were together and he was brilliant and happy!

As for myself, in 1974 I was elected corresponding member, and then full member, of the Accademia dei Lincei. Now I am close to being pensioned; I hope, however, still to be able to do something useful, because I feel that to clarify ideas and theses in probability is never enough.

3. One, No One, A Hundred Thousand

"Everybody speaks about Probability, but no one is able to explain clearly to others what meaning probability has according to his own conception."

This remark (by Garret Birkhoff) was made rather a long time ago, but the strange situation does not seem to have changed substantially to this day. Many different and contradictory definitions, or pseudo-definitions, several interpretations and ways of applying probabilistic reasoning have been suggested and used; one might, in fact, be led to use the title of a well-known novel by the Italian author Luigi Pirandello, *One*, *No One*, *A Hundred Thousand*, to describe the position.

The only approach which, as I have always maintained and have confirmed by experience and comparison, leads to the removal of such ambiguities, is that of probability considered—always and exclusively—in its natural meaning of "degree of belief." This is precisely the notion held by every uncontaminated "man in the street."

4. A Significant Anecdote

Let me relate an unexpected confirmation of the correct understanding of probability by laymen. I once asked a barman in Rome about the meaning of the sequences of numbers, such as 50-30-20, or 25-60-15, displayed in tabular form in the window of the café, and referring to the football matches for the forthcoming Sunday. "They are the probabilities!" he exclaimed in surprise: no doubt he thought he must have met the only person in the world ignorant of the meaning of "probability." And I was happy, because this was a confirmation of my opinion that probability is well understood and correctly interpreted by everybody, except perhaps the wiseacres who distort its meaning for the idle purpose of transforming it into something nobler than it is. But nothing becomes nobler through disguise: quite the contrary!

Unfortunately, most statisticians and philosophers (or pseudo-philosophers) endeavour to disguise the true nature of probability with this impossible aim. Or, as Hans Freudenthal has suggested, they hesitate to present probability "as God created it," because of the risk of being incriminated in an offense against decency.

This remark strictly agrees with a saying of Poincaré's to the effect that "la théorie des probabilités n'est que le bon sens réduit au calcul." If one forgets or neglects such an essential recommendation, or makes use of poor ad hoc methods or pretentious complications, one risks talking nonsense.

5. Promoting "Probabilistic Football Forecasts"

I am interested in prediction in appropriate fields. Football forecasting, although it has no truly practical applications, is a very good training ground for assessing the probabilities of all kind of events.[1]

The application of these probabilistic forecasting methods is not peculiar to football matches. Such forecasts may properly be applied to all kinds of uncertain events: their application to football is interesting simply because everyone has access to the same amount of information from past records, and news about the teams from the newspapers.

"Probabilistic forecasting" is a completely different kind of competition from the usual ones, which consist of an attempt to predict football results. It is not based on betting on a given event: here the results "win," "lose" or "draw" are denoted by the usual symbols 1, 2, X.

Probabilistic forecasting consists in providing the three probabilities p_1, p_2, p_X (e.g. 50-30-20) which can be set out as perpendicular distances from a point within an equilateral triangle to each of its three sides (see Figure 1). The simplest and most commonly applied of the "proper scoring rules" is "Brier's rule." It consists in using as penalty (or "negative points") the squares of the distances between the "predicted point" and the point corresponding to the true result.

[1] I consider it an important educational goal to learn to estimate numerically any measure such as probability, distance, weight, temperature, and so on. Think, for example, of the education of Rudyard Kipling's Kim to enable him to gather information for the English authorities in colonial India.

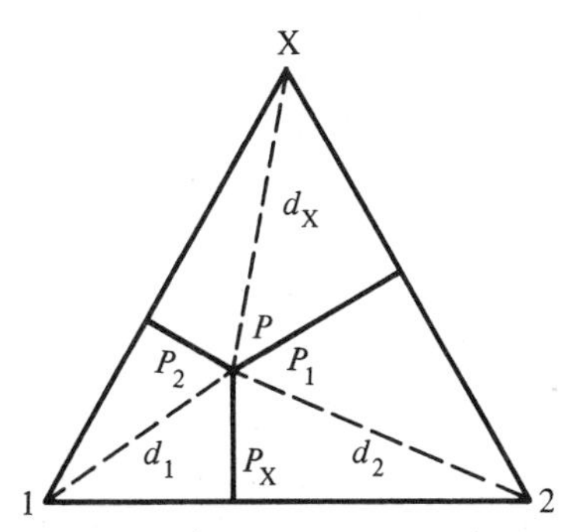

Figure 1. Probabilistic forecasting diagram. $P_i = \Pr\{i\}(i=1,2,X)$. $d_i =$ distance from P to vertex i. $d_i^2 =$ penalty if true result is i.

The classification for any single week is therefore given by the square of the distance between the "predicted point" and the vertex 1, 2 or X corresponding to the true result of a match. The progressive classification is obviously given by the total of all penalties from the start of the competition to the present.

The calculations, carried out by IBM computers, were completed every Sunday and Monday. Some data from the printouts are shown in Tables 1 and 2.

Regrettably, all this has been interrupted because of a period of disorders and strikes both at the universities and in the postal services in Italy. This is

Table 1. Football Forecasting Competition, 1970–71. Partial Listing of the Results Based on Matches Played on 14 March 1971, the 19th Week of the Season

Identification number	Name	This week's position	This week's penalty points	Previous position	Present position	Total penalty points
151	Boni	10	1.810	1	1	41.309
20	Rizzi	8	1.787	2	2	42.514
13	Valentini	42	2.157	3	3	42.950
10	de Marchis	41	2.157	4	4	43.129
15	de Paola	43	2.157	5	5	43.362
11	de Marchis	36	2.107	6	6	43.647
155	Sirotti	17	1.887	8	7	44.126
14	Campoli	38	2.107	7	8	44.140
24	Costanza	15	1.870	9	9	44.362
30	Marini	9	1.800	12	10	44.451
142	Dannucci	5	1.702	14	11	44.509
31	Caranti	21	1.932	10	12	44.539
12	Luciani	37	2.107	11	13	44.724
150	de Finetti	44	2.210	13	14	44.923

Table 2. My Own Weekly Forecasting Sheet, 14 March 1971

Match		Result	My own forecast	Penalty points	Published forecast	Penalty points
Bol	Var	1	55-35-10	.167	59-35-16	.155
Cag	Juv	X	25-40-35	.272	32-39-39	.314
Cat	Fog	1	25-50-25	.437	35-37-37	.348
Mil	Fior	1	50-40-10	.210	63-29-17	.125
Nap	Lrv	1	50-35-15	.197	66-27-16	.108
Roma	Laz	X	40-40-20	.280	51-35-23	.371
Tor	Int	2	25-40-35	.322	25-33-52	.200
Ver	Samp	1	35-40-25	.322	33-45-32	.379
			Total penalty points	2.210	Total penalty points	2.000

a pity, and I should like to see someone start up this simple and instructively amusing venture again, and keep me informed about his experience.

6. Grayson's Work on Drilling Decisions

The most enlightening example I know of proper probabilistic reasoning and the mathematical developments of the theory of probability with reference to practical but not standard problems (chiefly concerned with decision making) may be found in a book by C. Jackson Grayson, Jr. [12].

I think it advisable, when teaching probability, always to refer to real problems and situations. I prefer to single out well-specified cases, rather than submerging any single case of interest in a collectivity of more or less similar cases, which is quite the wrong approach. For example, insurance companies use standard mortality tables, but only as a rough initial reference, to be altered according to the more or less favourable results of a medical inspection, and other personal circumstances such as a dangerous life style.

Contrary to the widely accepted opinion, frequency and probability are completely different concepts, indirectly related by the fact that the expected relative frequency is the average of the probabilities of the single events envisaged, and the probability may be (under ordinary assumptions, but not necessarily) expected to be roughly close to the relative frequency.

Going back to Grayson, one of the most remarkable aspects of his methods is the systematic substitution of a numerical estimate of the probabilities concerned with a proposed drilling operation for the somewhat ambiguous "evaluative phrases." Several examples of such phrases are given in Grayson's book. The most remarkable of all is the systematic substitution of a numerical estimate of the probabilities for all ambiguous statements. Such statements can be interpreted either in an optimistic or in a pessimistic light, like the Sybil's phrases: for example "Ibis, redibis/non/morieris in bello," where a comma before or after the "non" suffices to reverse the prophecy. It is precisely in such cases that the estimated probabilities are most useful.

7. The Subjectivistic Interpretation of Probability

Subjective probability is the only interpretation, among the many different ones, that I consider meaningful and reasonable. It is, as I have already stated, the one that interprets probability in its natural meaning of "degree of belief."

Such a definition may be interpreted in the sense of betting behaviour: the probability of the event E is then 64 per cent, say, if one asserts that for a fair bet on E, one should pay \$64 to win \$100 if E occurs.

This procedure is, however, asymmetrical: the bettors may have different amounts of information. A neutral way of "betting" may be more desirable; that is, one based on proper scoring rules. Let us consider only the simplest of these: "Brier's rule" based on quadratic penalties. If any bettor indicates what is, in his opinion, the probability, say p, of the event envisaged, then the penalty is $(E-p)^2$, that is $(0-p)^2=p^2$ if $E=0$ (event does not occur) and $(1-p)^2$ if $E=1$ (event occurs).

We have identified the event E with the number 1 or 0 according to whether E occurs or not. In an unnecessarily complicated notation and nomenclature, the event E (if, as here, it is identified with the "indicator" $E=1$ or $E=0$ according by whether E occurs or not) is called "the indicator of E." But, as a Latin saying recommends: *entia non sunt multiplicanda sine necessitate* ("Don't multiply entities unnecessarily").

If any participant in the competition indicates what is, in his opinion, the probability, say p, of the event E envisaged, the penalty is $(E-p)^2$. The progressive sums of the penalties will lead to a final classification, and the amount of the bets paid to participate in the betting experiment will be returned to bettors in inverse proportion to the penalties they have accumulated.

I have referred above to competitions of this kind for football, which were carried out for several years at the University of Rome. I hope such experiments will continue, if only to discourage rougher methods of betting. Proper scoring rules require careful assessments of probability; betting invites just the contrary. Bettors are willing to risk very strange predictions so that, if by an extraordinary chance they are successful, they will become among the few very rich men in the world. In practice, what will almost certainly happen, is the defeat of anyone trying to win en plein with unlikely *predictions* (100-00-00; or 00-100-00; or 00-00-100) instead of true *previsions*.

The reader will by now have understood why the notion of "objective probability" and all related ones are meaningless, or in some way inconsistent. But there exists a very large field of strange ideas and terminologies as well as types of problems, often tremendously complicated and far from suitable for concrete applications or for meaningful discussion. These unfortunately occupy a large proportion of statisticians' time today. The worst of all attempts to rebuild the tower of Babel is the one "defining" (!) the probability of an "event" as the limit of the relative frequency of "trials of such an event!" But, as if that were not strange enough, one is also asked to accept the assertion that all the infinitely many "trials" are "independent," "equally likely" and "random." This last requirement is called an axiom; obviously, such an axiom can have no possible verification until the end of mankind's survival, or until the collapse of our universe.

I believe all this is idle self-deceit. Even if the time were limited to the lifetime of our galaxy, the limit could not be attained, and such wordings can be nothing but trompe-l'oeil.

There are many other pseudo-conceptions of a similar kind. Among those who amuse themselves by speaking of imaginary denumerable sequences of "events," some add the condition of irregularity (called the *Regellosigkeits-axiom* by R. von Mises and others). This is intended to impose the condition that even in eternity no rule should be possible. But of what interest is such a hypothesis, if, "in the long run," nobody will survive to ascertain whether it is right or wrong?

I consider such questions completely idle. I have felt obliged to say something on this topic in order to express my opinion that there is little reason to work on such pseudo-problems. Many other examples of "regularity" and "irregularity" and similar topics could easily be added. But I believe there is little interest in considering such theories; moreover, I would not wish to condemn totally the ideas of respectable scientists who see these things from a different point of view.

8. The Need for Unequivocal Terminology

The need for unequivocal terminology is obviously essential, particularly in the fields of probability and statistics, which are not carefully distinguished one from the other; confusion on this point can be terribly misleading.

First of all, we must clearly distinguish statistics, which deals with objective data and classifications in conditions of certainty, from probabilistic data which expresses somebody's degree of belief (or the average degree of belief of some individuals), about some events.

I must particularly insist on noting ambiguous wordings, especially about the term "event." Statisticians always use the term "event" to indicate a whole set of events (more or less clearly described or delimited). It seems to me preferable to call each single case an "event" instead of the usual single "trial of an event."

It would take too long to consider all the more or less improper wordings and notations. As an example, let me point out that "independent equally likely events with the same (but unknown) probability" is contradictory. In such a situation the correct statement requires a new category, that of "exchangeable events." Fortunately, "exchangeability" now seems rather well accepted. If the composition of an urn containing colored balls is not known, the result of each "trial" is informative, so that the opinion about its composition will be modified in the direction of the color drawn. But it seems that only a few years ago this circumstance was overlooked.

The only possible—but impractical and rather confusing—interpretation of the wording "objective probability" would be that the probability of a sure event which we know to have occurred is 1, and it is 0 otherwise. Also in the general case, it may be said, humorously but correctly, that the "true probability" is its "truth value": 1 if E does occur and 0 if it does not. In the case of uncertainty, the probability $p = P(E)$ is the average of 0 and 1 with the corresponding weights p and $(1-p)$.

As for my rejection of the wording "equally likely and independent events with unknown probability," and its replacement by "exchangeable events" I should remark that, in order to have "dependence" it is not necessary that the composition be changed, but simply that our state of information change. This is why it is necessary, in such cases, to speak of "exchangeable" events, not of "independent" ones.

9. My Pedagogic and Didactic Interests

I am very deeply interested in teaching: a field where I particularly value a vivid and imaginative presentation with many practical (and if possible amusing) applications. Not only for children, but for everybody.

The axiomatization and complex abstract presentation of mathematical notions and methods should, in my opinion, be a valid integration and final crowning of what has been carefully studied and understood in a more practical, intuitive and vivid manner. Not vice versa!

It is in this spirit that I wrote my papers (about 200) and several books, of which the main ones are listed below. I should mention that [8] was the fruit of discussions with Jimmie Savage when he was in Italy, and of later correspondence with him; he made several transatlantic suggestions.

For younger students, I wrote [5] for didactic purposes. With the same aim in mind, I organized, with the cooperation of some university colleagues, mathematical lectures on interesting and amusing subjects, and mathematical competitions for students in high schools. Unfortunately once again the period of disorders and other difficulties has obliged me to interrupt such useful and exciting activities; but I wish to remain active in these fields and hope that the interruption is only temporary.

Publications and References

[1] DE FINETTI, B. (1943) *Matematica logico-intuitiva*. E.S.T., Trieste. (Second edition 1959, Cremonese, Roma.)

[2] DE FINETTI, B. (1956) *Lezioni di matematica finanziaria*. Ricerche, Rome.

[3] DE FINETTI, B. (1957) *Lezioni di matematica attuariale*. Ricerche, Rome.

[4] DE FINETTI, B. (1957) *Economia delle assicurazioni*, I. UTET, Torino.

[5] DE FINETTI, B. (1967) *Il 'saper vedere' in matematica*. Loescher, Torino. (Second edition 1974; German translation, *Wie sucht man die Lösung, Mathematischer Aufgaben*, Birkhauser, Basle.)

[6] DE FINETTI, B. (1969) *Un matematico e l'economia*. F. Angeli, Milan.

[7] DE FINETTI, B. (1970) *Teoria delle probabilità*. Einaudi, Torino. (English translation, *Theory of Probability*, Wiley, London, 1974–1975; German translation, *Wahrscheinlichkeitstheorie*, Oldenbourg, München, 1981.)

[8] DE FINETTI, B. (1972) *Probability, Induction and Statistics*. Wiley, London.

[9] DE FINETTI, B. (1973) *Requisiti per un sistema economico, etc.* F. Angeli, Milan.

[10] DE FINETTI, B. (1976) *Dall'Utopia all'alternativa*. F. Angeli, Milan.

[11] DE FINETTI, B. and MINISOLA, F. (1961) *La matematica per le applicazioni economiche*. Cremonese, Rome.

[12] GRAYSON, C. J. JR. (1960) *Decisions under Uncertainty: Drilling Decisions by Oil and Gas Operators*. Harvard Business School, Division of Research.

Eugene Lukacs

Professor Eugene Lukacs was born in Szombathely, Hungary, on 14 August 1906. He was brought up in Vienna and completed his secondary schooling at a Realgymnasium there in 1925. He first decided to study mechanical engineering at the Technical University, but later transferred to the University of Vienna to read mathematics.

After graduating in 1930, he took an actuarial degree in 1931. He taught for two years at secondary schools before accepting a position in the mathematics division of an insurance company. When Germany annexed Austria in 1938, he decided to emigrate to the USA, where he arrived in February 1939.

Under the influence of Abraham Wald, he became interested in statistics and began his academic career in the USA at Illinois College, Berea College, and Our Lady of Cincinnati College. He worked for the Office of Naval Research in 1953–4 and became a member of the Mathematics Department at the Catholic University of America in Washington in 1955, retiring from it in 1972. He then joined the Bowling Green State University, Ohio, from which he retired in 1975.

Professor Lukacs has received many honours during his professional life: he is a Fellow of both the Institute of Mathematical Statistics and the American Statistical Association. He was elected to membership of the International Statistical Institute in 1963 and the Austrian Academy of Sciences in 1973. He has fulfilled several important editorial duties, including that of being editor of the Academic Press series in Probability and Mathematical Statistics. He has published several books and a large number of papers.

He married Elizabeth C. Weisz in 1935; he enjoys research and travel.

From Riemannian Spaces to Characteristic Functions: The Evolution of a Statistician

Eugene Lukacs

1. My Early Life and Education

I was born in Szombathely, Hungary on 14 August 1906. When I was six weeks old I was brought to Vienna, Austria where my parents had established residence several years earlier. I was educated bilingually in Hungarian and German, and spent a substantial part of each summer in Hungary. I had friends and playmates in both countries.

I went to elementary school in Vienna, but during the summer I was taught by an Hungarian teacher so that I spoke and wrote Hungarian and German with equal facility. Having finished the elementary school I entered a Gymnasium, where Latin was taught from the first year on and Greek from the third year on. After the end of the third year I decided that I did not want to study Greek and transferred to a Realgymnasium where French was taught instead of Greek and where mathematics and descriptive geometry were emphasized. I graduated from this school in 1925 with the degree Matura.

My interest in mathematics developed during my last three years at the Realgymnasium. Nevertheless I decided to study mechanical engineering since mathematics was considered an "impractical" subject with hardly any job opportunities. I registered at the Technical University but, apart from the required engineering courses, I also took courses in projective and descriptive geometry and audited some mathematics courses at the University of Vienna. After finishing the first year at the Technical University I decided to transfer to the University and to study mathematics. I took courses with Professors Hans Hahn, Wilhelm Wirtinger, Leopold Vietoris, Edward Helly and Walter Mayer.

I was at this time still very interested in geometry, and so I wrote my doctoral thesis under W. Mayer on "Planes in Riemannian Spaces" and obtained my Ph.D. on 4 July 1930. In November 1929 I had obtained a secondary school teaching diploma (Lehramtsprüfung) with descriptive geometry and mathematics as my subjects. I subsequently took the (academic) degree in actuarial science in March 1931.

2. Teaching and Actuarial Work in Vienna

After teaching for two years at secondary schools in Vienna, I accepted a position with the Phoenix Life Insurance Company in August 1931. The atmosphere in the mathematics division of Phoenix was almost academic. The chief mathematician was a university professor and the staff consisted of Edward Helly (also of the university) and Z. W. Birnbaum who later became a professor at the University of Washington in Seattle.[1] In addition there were several active mathematicians in other divisions.

It was during my time at Phoenix that I married Elizabeth C. Weisz (Lisl) on 20 January 1935. We had met in 1929 at the university where she studied mathematics and physics. Lisl has always taken a great interest in my professional activities and has helped me wherever she could, although she had full-time teaching positions either at high schools or colleges. I am grateful for her help during all our years together.

When the Phoenix company failed, I was employed by its successor, the Övag company, until April 1937. Subsequently I worked (until April 1938) at the Assicurazioni Generali di Trieste where I dealt with non-life insurance. From 1928 until 1938 I taught mathematics at the Volkshochschule Wien Volksheim. This institution offered university extension courses; I enjoyed this work in adult education very much. The annexation of Austria by Nazi Germany in 1938 terminated my professional career in Vienna and I emigrated to the USA, arriving there in February 1939.

3. A New Life in the USA

Soon after my arrival in the USA I met Abraham Wald whom I had previously known in Vienna. Wald had obtained a degree (roughly equivalent to an MA) at the university of Cluj, Romania, and then moved to Vienna in order to continue his studies in mathematics. I had met him at the University of Vienna soon after his arrival; he joined a group with which my wife and I used to hike in the Vienna Woods.

This was a group of young mathematicians and we often discussed problems in mathematics or philosophy during our hikes. Wald was a fine addition to the group, with his pleasant and modest personality and his varied interests and deep thoughts. Some work which Wald did in the axiomatics of geometry clearly showed his great talent, and he continued for some time to work primarily in geometry. But while he was still in Vienna he had begun to interest himself in probability and statistics. When I met him again in the USA, Wald drew my attention to the vast and highly significant amount of work in mathematical statistics and probability theory

[1]See Professor Birnbaum's paper in this volume, pp. 75–82.

which was largely unknown in Central Europe at that time. He suggested that I familiarize myself with this work and invited me to attend informally his own and Hotelling's courses at Columbia University, New York.

My interest has in recent years been focused on the theory of characteristic functions. The contribution of Soviet mathematicians to this important chapter of probability is considerable; take for example stability theory, which interests me very much. Three special meetings, held recently in the USSR, have been devoted to stability theory; the results of these conferences were published but are little known in the USA. I have also done some work on stability and have tried to bring this area to the attention of my colleagues.

I arrived in the USA during the aftermath of the great depression, so that it was impossible for me to find a suitable position. I first obtained a teaching job at a high school and later I held posts at several colleges; finally I accepted a professorship at Our Lady of Cincinnati College in Cincinnati, Ohio, a position I held from September 1945 until June 1953. There I taught mathematics, but was on leave during the academic years 1948 to 1952. During the war years, I taught physics and had no chance to do mathematics. In Cincinnati I contacted Otto Szász, whom I had already met at the Hellys in Vienna. Szász stimulated me to resume working in mathematics. He gradually became interested in my probability problems, and this resulted in several joint papers. During my leave from Our Lady of Cincinnati College, I was employed as a mathematical statistician first at the U.S. Naval Ordnance Test Station in China Lake, California from September 1948 to May 1950 and then from September 1950 until October 1953 at the National Bureau of Standards in Washington, DC.

4. The Statistical Laboratory, Catholic University of America

I transferred to the Office of Naval Research in October 1953 and became head of its Statistics Branch. During my stay at the Bureau of Standards and at the Office of Naval Research I also taught at the American University in Washington. I eventually joined the Catholic University of America in Washington on 1 September 1955 and retired from it in May 1972 with the title of Professor Emeritus. At the Catholic University I organized the Statistical Laboratory as an autonomous group within the Mathematics Department in 1959, and was its director until my retirement. Among the members of the Statistical Laboratory were Professor E. Batschelet, T. Kawata, R. G. Laha, M. Masuyama and V. K. Rohatgi. Batschelet worked in biometry; the interests of Kawata and Laha were close to my own in analytic probability theory and I cooperated with Laha for some time. Kawata wrote a book while in Washington; Masuyama was interested in the design of experiments; and Rohatgi worked in mathematical statistics.

I organized lecture series in probability and statistics; lectures were given every month by prominent workers in the field.

At my request the university agreed to establish a visiting professorship in the Statistical Laboratory. This position was filled by outstanding foreign scientists like H. Cramér and P. Lévy, who usually stayed for one academic year, and occasionally for only one term. The visiting professorship, as well as the lecture series, proved very stimulating for the faculty as well as for the students of the Statistical Laboratory.

By 1972 Professors Kawata and Masuyama had been forced to return to Japan for family reasons. The financial situation of the university had by that time deteriorated to such an extent that it was not possible to replace them. This made it clear to the remaining staff members of the Statistical Laboratory that it was impossible to continue work at the level they considered desirable. Professor Batschelet received a call to the University of Zürich, Switzerland. I was offered a position at Bowling Green State University in Ohio, and suggested that similar offers be extended to Professors Laha and Rohatgi. When these offers materialized, Laha, Rohatgi and myself accepted them; we resigned from the Catholic University at the end of the term and moved to Bowling Green. This terminated the work of the Statistical Laboratory and necessitated its dissolution.

5. Bowling Green State University and After; My Travels

I became a professor of mathematics at Bowling Green State University, Ohio, in September 1972. In 1976 its Board of Trustees honored me with the title of University Professor. I retired from this position on 30 September 1976. But this did not mean the end of my work. I was a visiting professor at the Institute of Technology in Vienna and taught there during the academic years 1975–76 and 1976–77. During the academic year 1977–78 I was a visiting professor at the University of Erlangen and also had the opportunity to visit other German universities. I returned to Washington, DC in the late fall of 1978 and taught one course at Catholic University during the academic year 1979–80.

Throughout my career, I have always enjoyed travel and visits to overseas centers. During a sabbatical leave from Catholic University in the academic year 1961–62, I spent the first term at the Sorbonne where I taught two courses. During the second term of my sabbatical I was in Zürich at the Swiss Federal Institute of Technology (ETH) where I taught one course and gave one seminar. While in Paris I had the opportunity of meeting Paul Lévy, one of the leading probabilists of our time. Subsequent visits to Paris strengthened my contact with Professor Lévy who was later one of the visiting professors at the Statistical Laboratory of the Catholic University. Paul Lévy was one of the most prominent probabilists of this period and it was very interesting to listen to him. I also contacted the eminent French mathematicians Professors Fréchet and Darmois, whom I had met at an earlier meeting in Belaggio, and established a close personal relationship with D. Dugué; we had many common interests.

During 1961–62 I also lectured at the Université Libre de Bruxelles and at the University of Athens. I was again on leave during the academic year 1965–66 and spent five months in Vienna and the rest of the time in Paris at the Sorbonne. I was also on leave from Catholic University from 1 February 1970 to August 1970 and was a visiting professor at the Institute of Technology in Vienna during this period. I spent the month of May 1971 with my colleague Toby Lewis at the University of Hull, England. During part of the academic year 1974–75 I was a senior visiting fellow in Joe Gani's Department of Probability and Statistics in Sheffield, England.

6. Fifty Years of Research and Professional Activities

In May 1980 I received the golden doctoral diploma of the University of Vienna. This honor is awarded to graduates on the fiftieth anniversary of their Ph.D. provided that they have continued their research work since receiving their degree.

During my frequent stays in Vienna I also had the opportunity to visit the Mathematical Research Institute of the Hungarian Academy of Sciences in Budapest. Most recently I accepted an invitation to Brazil for the fall of 1980 and lectured for two weeks at the University of Ribeirao Preto and for two weeks at the University of Campinas.

During the years I spent in Washington and Bowling Green, I travelled and lectured a good deal at conferences and international meetings. I attended the International Mathematical Congresses in Oslo (1936), Cambridge (1950), Amsterdam (1954), Edinburgh (1958), Stockholm (1962), Moscow (1966), Nice (1970), Vancouver (1974), and Helsinki (1978). I also participated in the sessions of the International Statistical Institute in Rome (1953), Petropolis (1955), Stockholm (1957), Brussels (1958), Tokyo (1959), Paris (1961), Ottawa (1963), Belgrade (1965), Sydney (1967), London (1969), Washington (1971), Vienna (1973), Warsaw (1975) and New Delhi (1977). I participated in four Prague Conferences and six Berkeley Symposia. At all these international gatherings I gave either papers or invited addresses.

During the summer of 1980 I organized, jointly with D. Dugué, a conference at the Mathematical Research Institute in Oberwolfach on analytical methods in probability theory. The proceedings of this conference have now been published by Springer-Verlag.

Early in my career I joined the American Mathematical Society (AMS), the Institute of Mathematical Statistics (IMS), whose activities were close to my personal interests, the American Statistical Association (ASA), the Biometric Society of Switzerland, the Mathematical Society of Vienna (until 1938), and the Austrian Mathematical Society (after World War II). I was elected a Fellow of IMS in 1957, of the American Association for the Advancement of Science in 1958 and of the ASA in 1969. I was also elected to membership of the International Statistical Institute in 1963 and to the Austrian Academy of Sciences on 23 May 1973.

My editorial responsibilities have been very varied. I was a cooperating member of the editorial board of the *Annals of Mathematical Statistics* from 1958–1964 and an associate editor of the same *Annals* from 1968–1970. From 1951–1955 and 1961–1963 I was an editorial collaborator of the *Journal of the American Statistical Association*. I have been on the editorial board of the *Journal of Multivariate Analysis* since 1970. I am chairman of the IMS translation committee and as such a member of the AMS translation committee.

I remain a reviewer for *Mathematical Reviews* and for the *Zentralblatt für Mathematik* and have written reviews for other professional journals. Since 1962 I have been editor of the Probability and Mathematical Statistics series published by Academic Press.

7. Concluding Remarks

Despite my wanderings and some ill health in recent years, I have led a happy life shared by my wife Lisl. My work and my statistical and mathematical interests have sustained me through many difficulties. I still do some research and I hope that my books and papers have contributed something to the development of statistics in recent years.

My books, as well as my first three and my last three papers, are listed below for the reader's information; their titles demonstrate how my research interests have developed during a half-century of professional life.

Publications

[1] Bergmann, G. and Lukacs, E. (1930) Ebenen und Bewegungsgruppen in Riemannschen Räumen (Planes and groups of motion in Riemannian spaces). *Monatsh. Math. Phys.* **37**, 303–324.

[2] Lukacs, E. (1930) Anwendungen der mehrdimensionalen darstellenden Geometrie auf Nomographie (Applications of the descriptive geometry of several dimensions to nomography). *Z. Angew. Math. Mech.* **10**, 501–508.

[3] Lukacs, E. (1932) Über zwei theoretische Fragen der Nomographie (On two theoretical problems of nomography). *Z. Angew. Math. Mech.* **12**, 244–251.

[4] LUKACS, E. (1960) *Characteristic Functions*. Griffin, London. (Second edition 1970; third edition in preparation.)

[5] LUKACS, E. (1968) *Stochastic Convergence*. D. C. Heath, Boston. (Second edition, Academic Press, New York, 1975.)

[6] LUKACS, E. (1972) *Probability and Mathematical Statistics: An Introduction*. Academic Press, New York.

[7] LUKACS, E. (1979) Characterization of the rectangular distribution. *Stoch. Proc. Appl.* **9**, 273–279.

[8] LUKACS, E. (1979) Stability theorems for the characterization of the normal and of the degenerate distribution. In *Asymptotic Theory of Statistical Tests and Estimation*, Academic Press, New York, 205–230.

[9] LUKACS, E. (1980) Inequalities for Fourier transforms of functions of bounded variation. In *General Inequalities* II, ed. E. F. Beckenbach, Birkhauser-Verlag, Basel, 127–133.

Gheorghe Mihoc

Gheorghe Mihoc was born on 7 July 1906 in Brăila, Romania, the son of Gheorghe and Ecaterina Mihoc. After completing his secondary education at the Gheorghe Sinçai gymnasium in Bucharest, he studied mathematics at the University of Bucharest, and later at the University of Rome. He obtained his degree of Doctor of Statistical and Actuarial Sciences at the University of Rome in 1930, and four years later that of Doctor in Mathematics, at the University of Bucharest.

From 1937 to 1973, he held in succession the posts of assistant, lecturer, and professor at the University of Bucharest. Between 1929 and 1948 he was also working as an actuary in social insurance. Professor Mihoc has published a large number of research papers in probability theory, mathematical statistics and mathematical analysis. He has, however, made his name not only as a scholar, but also as an active worker for the development of mathematics and statistics in Romania. He has been Director General of the Central Board of Statistics (1948–51), Dean of the Faculty of Mathematics and Physics of Bucharest University (1951–59), and Rector of the University of Bucharest (1963–68).

Professor Mihoc held the post of Director of the Centre of Mathematical Statistics in Bucharest for twelve years between 1964 and 1976, and was elected Fellow of the Romanian Academy in 1955. He rose to be its President in 1980, a position he currently holds. Among his many other honours is the award of the First Class Labour Order of Romania.

He was married on 17 September 1953; Professor and Mrs. Mihoc have no children.

Regrettably, Professor Mihoc died on 25 December 1981 while this book was in proof.

A Life for Probability

Gheorghe Mihoc

1. Early Education

I was born on 7 July 1906 in Brăila, a picturesque port on the Danube. My parents had previously left Banat, their birthplace, to try to earn their living elsewhere. Circumstances drove them further, to Bucharest, where they settled in 1908; thus I was brought up, studied and have lived in Bucharest for more than seven decades.

As a student at the Gheorghe Şincai gymnasium, I had some good mathematics teachers: Ovidiu Ţino, Theodor Angheluţă, Nicolae Niculescu, all of whom inspired my love for this science. But there was still another element that directed me at an early stage towards mathematics, namely the existence of the *Mathematical Gazette*. This journal, founded in 1895 and still active today, proved to be a real focus of attraction for young people interested in mathematics. In its pages the names of the young authors who solved and proposed mathematical problems were mentioned. The fact that students have the opportunity to publish problems which they have discovered is of great pedagogical importance. During lessons, as well as in their private work based on textbooks, they learn to solve problems, and this helps them to understand more deeply what they are taught. The work of creating problems achieves a similar goal, but adds inventiveness and imagination, whose roles in mathematics are well known. The activities of the *Mathematical Gazette* provided a very useful apprenticeship for me and many other Romanian mathematicians.

2. The University of Bucharest

After graduating from the gymnasium, I enrolled in the mathematical section of the Faculty of Sciences in Bucharest. My studies there lasted from 1925 until 1928. The courses extended over three years and the amount of knowledge conveyed to us through lectures and seminars was rather restricted, although some of our teachers were of undeniable quality.

Our university training was uneven. From Professor Gheorghe Ţiţeica, a renowned geometer and a first-class teacher, we acquired an up-to-date knowledge of geometry. Unfortunately, the same could not be said about mathematical analysis where the lectures were outdated and had little connection with the important discoveries made in this field in the first decades of our century. I tried to fill this gap by my own efforts but I was nevertheless strongly affected by it when writing my first papers. In my last year, Professor Octav Onicescu gave us a course of lectures on probability. It was an inspired course outlining the state of the field at that time. This was my first contact with the discipline that was later to become my speciality.

I was quite exhausted on graduating, because of the efforts I had made to earn a living during my university studies. But my financially strained circumstances were not over with graduation. I could not find any work as a teacher: help came eventually from Professor Onicescu, who among our masters remained the closest to young mathematicians. He found me a job with the social insurance service, in the department of statistics and actuarial studies. As if this were not enough, pointing out that insurance calculations lacked an appropriate scientific background, Professor Onicescu convinced the leadership of the institute to grant me paid leave to specialize in actuarial mathematics in Rome.

3. Actuarial Studies in Rome

The choice of Italy was justified by the existence at the University of Rome of a recently founded Faculty of Statistical and Actuarial Sciences. In November 1929 I became a student of this faculty. The year 1929–30 spent in Italy was the happiest of my life; with nothing else to worry about, I could devote my time entirely to study. The professors in the Faculty of Statistical and Actuarial Sciences in Rome were a select group. They numbered among others Guido Castelnuovo, Francesco P. Cantelli, Corrado Gini, Rodolfo Benini, Franco Savorgnan, all of them great names in Italian science.

Castelnuovo is well known as one of Italy's greatest mathematicians: the Mathematical Institute in Rome is named after him. His papers made him famous in algebraic geometry. Although he specialized in geometry, Castelnuovo had a wide knowledge of all branches of mathematics, and lectured on subjects that were outside his speciality. One of these was probability theory, on which he delivered a masterly course published in two volumes which were considered authoritative for the period. Those who have a keen interest in the history of mathematics and are interested in knowing the state of probability theory prior to the 1940s, when its fusion with measure theory emerged, should read this book. It will hold their attention even today, by its clear style and its scientific distinction. When I had him as a teacher, Castelnuovo was around 70, a venerable old man with a white beard. Meticulous and rigorous when lecturing, he never underestimated the intuitive and philosophical approach, arousing the enthusiasm of his audience.

Francesco P. Cantelli was quite a different character. Highly strung, even irascible, always nursing a cigarette at the corner of his mouth, he was in love with his profession of actuary and anxious to enrich its mathematical content. He founded and directed the *Giornale dell'Istituto Italiano degli Attuari*, to which many of the greatest statisticians and probabilists of the time contributed. Nowadays he is remembered in probability theory mainly because of the Borel–Cantelli theorem.

Corrado Gini and Franco Savorgnan were statisticians who distinguished themselves by applying statistical methods to important social and demographic studies. The same was done by R. Benini in economics.

The Faculty of Statistical and Actuarial Sciences in Rome opened up a new world for me, one in which I learned about the innumerably varied applications which a mathematical discipline can have. Paradoxically, this discipline turned out to be one in which the results are not expressed with absolute precision, but with uncertainty. I chose Guido Castelnuovo as supervisor for my Ph.D. thesis. He advised me to study the appendix of Markov's 1912 book on probability [7], in which the celebrated mathematician treated the sequences of random variables which he had introduced, and which had brought him well-deserved glory. He also drew my attention to a paper by Serge Bernstein [1], in *Mathematische Annalen*, suggesting that I should apply the general theorems stated there to the particular case of Markov chains. In Markov's book it was shown that the limits of partial sums of variables forming a simple homogeneous finite Markov chain follow a normal law. In the Ph.D. thesis which I defended at the University of Rome in July 1930 I showed that this property of simple chains also holds for multiple chains. Using the theorems given by Bernstein, I also proved that the central limit theorem is valid for non-homogeneous multiple chains.

4. Early Research in Bucharest

When I returned to Bucharest, I continued work as an actuary, an employment which I thought suitable for someone specialized in probability theory and statistics. Very often when applying mathematics in practice, various problems emerge which are the starting points for new theories. At the same time I continued my personal research in probability theory. In 1934 I defended my Ph.D. at the University of Bucharest, my thesis [8] being devoted to multiple finite Markov chains. Let me give a few details of this problem here.

A multiple l-order (or l-dependent) Markov chain x_n, $n \in N^* = \{1,2,\dots\}$ is characterized by the property

$$P(x_n = \xi_n | x_{n-1} = \xi_{n-1}, x_{n-2} = \xi_{n-2}, \dots, x_{n-l} = \xi_{n-l}, \dots, x_1 = \xi_1)$$
$$= P(x_n = \xi_n | x_{n-1} = \xi_{n-1}, x_{n-2} = \xi_{n-2}, \dots, x_{n-l} = \xi_{n-l}) \qquad (1)$$

for any $1 \leq \xi_i \leq m$, $1 \leq i \leq n$, $n \in N^*$.

Consequently, if one considers the entire past up to time n

$$\xi_1, \dots, \xi_{n-1},$$

the hereditary property of the Markov chain is concentrated only in the l ancestors

$$\xi_{n-l}, \xi_{n-l+1}, \dots, \xi_{n-1}.$$

Denoting by

$$\varphi(\xi_n; \xi_{n-1}, \dots, \xi_{n-l}) \qquad (2)$$

the probability on the right-hand side of (1), I called this a descending fundamental probability.

In my thesis I found a very simple property, giving a more complete idea of the nature of the dependence generated by Markov chains; this was the time-reversibility of the Markov property. Assuming (1) to hold it follows that

$$P(x_n = \xi_n | x_{n+1} = \xi_{n+1}, x_{n+2} = \xi_{n+2}, \dots, x_{n+l} = \xi_{n+l}, x_{n+l+1} = \xi_{n+l+1}, \dots)$$
$$= P(x_n = \xi_n | x_{n+1} = \xi_{n+1}, x_{n+2} = \xi_{n+2}, \dots, x_{n+l} = \xi_{n+l}).$$

Consequently, if the Markov property holds in the descending direction it will also hold in the ascending direction, the order of multiplicity being the same in both directions. Therefore there exist ascending fundamental probabilities as well.

Let us consider the probability

$$P(x_n = \xi_n | x_\alpha = \xi_\alpha, x_\beta = \xi_\beta, \dots, x_\lambda = \xi_\lambda) \qquad n > \alpha > \beta > \cdots > \lambda. \qquad (3)$$

It is possible to show that any conditional probability (3) is a weighted average of fundamental probabilities (2). It follows that any conditional probability of x_n given the values taken by a group of variables $(x_\alpha, x_\beta, \dots, x_\lambda)$ with $\alpha, \beta, \dots, \lambda$ smaller than n, lies between the smallest and the greatest of the probabilities (2).

The properties of the probabilities of a multiple chain suggested to me an interesting class of transformations with applications in the theory of differential equations.

In Euclidean space R_m let us consider the points

$$M_k = (x_1^{(k)}, x_2^{(k)}, \dots, x_m^{(k)}), \qquad k \in N^*,$$

and the transformation $T^r_{n,a}$ defined by

$$x_i^{(n+1)} = f_i\big(x_1^{(1)}, x_2^{(1)},\dots,x_m^{(1)}; x_1^{(2)}, x_2^{(2)},\dots,x_m^{(2)};\dots;x_1^{(n)}, x_2^{(n)},\dots,x_m^{(n)};$$
$$a_1, a_2,\dots,a_r\big) \qquad i=1,2,\dots,m, \tag{4}$$

where $a_1,\dots,a_r$ are r essential parameters, $a=(a_1,\dots,a_r)$.

Applying the transformation $T^r_{n,a}$, to the points $M_1, M_2,\dots,M_n$, we obtain the point M_{n+1}. To simplify the notation, we write equations (4) as

$$M_{n+1} = f(M_1, M_2,\dots,M_n; a). \tag{5}$$

Applying $T^r_{n,b}$ to the points $M_2, M_3,\dots,M_n, M_{n+1}$ we get

$$M_{n+2} = f(M_2, M_3,\dots,M_{n+1}; b). \tag{6}$$

We require the transformations $T^r_{n,a}$ to have the property that

$$M_{n+2} = f(M_1, M_2,\dots,M_n; c), \tag{7}$$

where c is a function of a and b only, i.e.

$$c_j = g_j(a,b), \qquad j=1,\dots,r. \tag{8}$$

Although the transformations $T^r_{n,a}$ do not form a group, they may be studied using the theory of groups, and have interesting applications in the theory of differential equations outlined below.

Consider the differential equation

$$\frac{dy}{dx} = f(x,y).$$

I. If $n>4$ then among the arbitrary solutions $y_1,\dots,y_n$ of the equation there is no relation of the form

$$y_n = \varphi(y_1, y_2,\dots,y_{n-1})$$

with φ a regular analytic function of the y_i, $1\leqslant i\leqslant n-1$.

II. If among three arbitrary solutions y_1, y_2, y_3 of the equation we have a relation of the form

$$y_3 = \varphi(y_1, y_2),$$

φ being a regular analytic function of y_1, y_2, then by means of a transformation $y=F(Y)$ the equation is reducible to a linear differential equation of the first order.

III. If among four arbitrary solutions y_1, y_2, y_3, y_4 of the equation we have a relation of the form

$$y_4 = \varphi(y_1, y_2, y_3),$$

with φ a regular analytic function of y_1, y_2, y_3, then the equation is reducible to a Riccati equation by means of a substitution $y=F(Y)$.

Property III sheds an interesting light on the classical property of the Riccati equation: among four arbitrary solutions there exists a constant anharmonic ratio.

It is interesting to note the connection leading from consideration of the conditional probabilities of a multiple Markov chain to such problems of analysis.

In the period I am referring to, namely the 1930s, probabilists established the basic properties of Markov chains. I therefore naturally went on studying these problems, which were both fashionable and interesting, and which represented the first step in an approach to dependent random variables. Let me relate some of my contributions of the time in the theory of Markov chains which I think most interesting.

Consider a simple Markov chain x_n, $n \in N^*$ with transition probabilities p_{ij} $(i, j = 1, 2, \ldots, m)$

$$p_{ij} \geqslant 0, \qquad \sum_{j=1}^{m} p_{ij} = 1.$$

Denote

$$p_{ij}^{(n)} = P(x_{n+1} = j \mid x_1 = i).$$

At the time, many probabilists had considered the asymptotic behaviour of the probabilities $p_{ij}^{(n)}$ and had established that under certain conditions

$$\lim_{n \to \infty} p_{ij}^{(n)} = \pi_j.$$

Consequently, in the long term, there exists independence of the initial conditions.

I also dealt with this problem and gave a very simple formula for the π_j. I was especially concerned with the asymptotic behaviour of the moments of the sum $S_n = f(x_1) + \cdots + f(x_n)$, where f is some real-valued function defined on the states of the chain. Let us write

$$\mathfrak{M}_n(p_i) = E(S_n)$$

under the assumption that the initial probabilities of x_1 taking the values $1, \ldots, m$ are respectively

$$p_{i1}, p_{i2}, \ldots, p_{im}.$$

Under the same hypothesis, write

$$\mathfrak{M}_n^r(p_i) = E(S_n - \mathfrak{M}_n(p_i))^r.$$

Consider now the characteristic equation

$$|p_{ij} - \lambda \delta_{ij}| = 0 \qquad (i, j = 1, \ldots, m),$$

where δ_{ij} is the Kronecker delta. If $\lambda = 1$ is a simple root of the characteristic equation, then

$$\frac{\mathfrak{M}_n^{2r}(p_k)}{n^r} \quad \text{and} \quad \frac{\mathfrak{M}_n^{2r+1}(p_k)}{n^r}$$

always have limits independent of k as $n \to \infty$.

The proof is based on the following idea: the successive moments satisfy the following recurrence relations

$$\mathfrak{M}_n^{\nu}(p_k)=\sum_i p_{ki}\mathfrak{M}_{n-1}^{\nu}(p_i)+C_{\nu}^1\sum_i p_{ki}\sigma_{ki}(n)\mathfrak{M}_{n-1}^{\nu-1}(p_i)+\cdots$$

where

$$\sigma_{ik}(n)=\mathfrak{M}_{n-1}(p_k)-\mathfrak{M}_n(p_i)+f(i).$$

These lead to a system of different equations of the form

$$\begin{aligned} &\chi_i(n)-\sum_k p_{ik}\chi_k(n-1)=\sigma_i(n)+\psi_i(n), \qquad (i=1,2,\ldots,m),\\ &p_{ik}\geqslant 0, \qquad \sum_k p_{ik}=1, \end{aligned} \tag{9}$$

where $\chi_k(n)$ is a polynomial of degree r in n and $\psi_i(n)$ a function such that

$$\lim_{n\to\infty}\psi_i(n)/n^r=0.$$

The system (9), without the right-hand side, had already been completely solved by Fréchet in his book [4]. The general solution can be obtained without difficulty.

In the same way we find that

$$\lim_{n\to\infty}\frac{\mathfrak{M}_n^{2r}(p_i)}{\left[2\mathfrak{M}_n^2(p_i)\right]^r}=\frac{(2r)!}{2^{2r}r!}=\pi^{-1/2}\int_{-\infty}^{\infty}x^{2r}e^{-x^2}\,dx$$

$$\lim_{n\to\infty}\frac{\mathfrak{M}_n^{2r+1}(p_i)}{\left[2\mathfrak{M}_n^2(p_i)\right]^r}=0=\pi^{-1/2}\int_{-\infty}^{\infty}x^{2r+1}e^{-x^2}\,dx.$$

The central limit theorem follows. In his book Fréchet called the above procedure 'The method of Schultz and Mihoc' because Schultz [29] had also applied it at almost the same time.

The question then arises as to the conditions under which the limiting distribution is not normal. This amounts to the fact that the sequence $\mathfrak{M}_n^2(p_i)$ remains bounded when n increases to ∞. I gave a necessary and sufficient condition for this to hold, and Doeblin subsequently rediscovered the same result.

5. Collaboration with Onicescu

After these papers were published, my collaboration with Professor O. Onicescu began. It was natural for this to occur, not only because Professor Onicescu was my teacher and my permanent supervisor, but also because we were at the time the only mathematicians dealing with probability in Romania. We started a systematic investigation of homogeneous finite Markov chains using the characteristic function. This is a simple method allowing a complete solution to the central limit problem for finite chains, and also providing good results for more general Markov chains.

Suppose a Markov chain x_n, $n \in N^*$ is given and let us denote by $F_k(n; t; \theta)$ the characteristic function of the sum $f(x_1)+\cdots+f(x_n)-n\theta$, where θ is a constant, under the hypothesis that the initial probabilities of x_1 taking the values $1,\ldots,m$ are respectively $p_{k1},\ldots,p_{km}$. These characteristic functions satisfy the system of difference equations

$$F_k(n; t; \theta)=\sum_{j=1}^{m} p_{kj}e^{(f(j)-\theta)it}F_j(n-1; t; \theta). \tag{10}$$

The solution can be written as

$$F_k(n; t; \theta)=\sum_{h=1}^{m}\sum_{j=1}^{s}\frac{1}{(m_j-1)!}\frac{d^{m_j-1}}{d\lambda^{m_j-1}}\left[\lambda^n\frac{G_{kh}(\lambda; t)}{\psi_j(\lambda)}\right]_{\lambda=\lambda_j}, \tag{11}$$

where

$$\left|p_{kj}e^{(f(j)-\theta)it}-\lambda\delta_{kj}\right|=(\lambda-\lambda_1)^{m_1}(\lambda-\lambda_2)^{m_2}\cdots(\lambda-\lambda_s)^{m_s}, \tag{12}$$

$$\psi_j(\lambda)=\frac{\left|p_{kj}e^{(f(j)-\theta)it}-\lambda\delta_{kj}\right|}{(\lambda-\lambda_j)^{m_j}},$$

$G_{kh}(\lambda; t)$ being the cofactor of the (h,k)th entry in the determinant (12). The explicit expression of the characteristic function allows the direct computation of its asymptotic value as well as of its moments of various order.

Equation (10) can be extended to more general chains. Thus, let us consider a simple homogeneous Markov chain taking on real values. Let us denote by $F(y|x)$ the distribution of x_n conditioned by $x_{n-1}=x$. We have

$$0\leq F(y|x)\leq 1, \qquad \int_R d_yF(y|x)=1,$$

$$F_n(x; t; \theta)=\int_R e^{(u-\theta)it}F_{n-1}(u; t; \theta)\,dF(u|x), \tag{13}$$

$F_n(x; t; \theta)$ being the characteristic function of the sum $x_1+\cdots+x_n-n\theta$ under the hypothesis that the distribution function of x_1 is $F(y|x)$, i.e.

$$F_1(x; t; \theta)=\int_R e^{(y-\theta)it}\,dF(y|x).$$

Let us introduce the function

$$F(x; t; \theta)=\sum_n \lambda^n F_n(x; t; \theta)$$

which satisfies the integral equation

$$F(x; t; \theta)=\lambda\int_R e^{(u-\theta)it}F(u; t; \theta)\,dF(u|x)+1.$$

We studied this equation [28], from which followed the asymptotic properties of the chain considered. In the particular case when the function $F(y|x)$ has a density, one obtains integral equations that can be studied by Fredholm's theory.

I subsequently applied the method of the characteristic function also for investigating other properties of Markov chains, for instance vectorial Markov variables [14], iterations [9], the extension of the Poisson law to Markov chains [10], [11], [12], and Markov chains with rewards ([25]; see pp. 332–357).

Professor Onicescu and I concerned ourselves with the problem of generalizing Markovian dependence. Let us again consider a simple homogeneous Markov chain x_n, $n \in N^*$. What is characteristic of this type of dependence is the fact that the distribution of x_n conditioned by $x_1 = \xi_1, \ldots, x_{n-1} = \xi_{n-1}$ is of the form

$$F(x; \xi_{n-1}).$$

In other words the conditional distribution depends only on ξ_{n-1}, although the entire past $\xi_1, \xi_2, \ldots, \xi_{n-1}$ is known.

Dependence with complete connections, as we defined it, links the variables x_{n-1} and x_n in a closer way. The variable x_{n-1} takes on the values ξ_{n-1} with certain probabilities depending in general on the preceding values $\xi_1, \ldots, \xi_{n-2}$. If these are known, let us denote by $F_{\xi_1 \cdots \xi_{n-2}}(x)$ the conditional distribution of the variable x_{n-1}. By definition, in the case of a simple chain with complete connections, the conditional distribution of x_n is of the form

$$F\left(x; \xi_{n-1}; F_{\xi_1, \ldots, \xi_{n-2}}(x)\right) \tag{14}$$

so that it is a functional of $F_{\xi_1, \ldots, \xi_{n-2}}(x)$ varying with ξ_{n-1}.

Let us examine the type of dependence given by (14). Let us call the conditional probability

$$P(x_n = \xi_n | x_{n-1} = \xi_{n-1}, \ldots, x_1 = \xi_1) \tag{15}$$

the effective probability of realization of the variable x_n. This effective probability is the conditional probability of realization of x_n, given the entire past as expressed by the values taken by the preceding variables. The connection between x_n and x_{n-1} is complete, in the sense that the effective probability of x_n depends on the two elements characterizing x_{n-1} as a random variable: its value and the effective probability with which this value is taken. In the case of Markovian dependence, the effective probability of x_n depends only on the value taken by the variable x_{n-1}.

6. Processes with Dependence: Complete Connections

We are naturally confronted with interdependences which lead to the consideration of natural processes where these might arise. Thus, if parents are consumptive, there is a higher probability of their children catching the same disease. For newborn children of healthy parentage, the probability of becoming consumptive is smaller than for the offspring of diseased parents. If we accept that such contagion depends only on whether the parents are or are not suffering from the disease, we get a Markov chain. If, however, we take into account, as seems natural, the parents' predisposition towards consumption, the dependence is more complex and the Markov chain is no longer adequate for a satisfactory description of the process. It is more logical to assume that the probability of an nth-generation individual's becoming consumptive depends on whether his parents were or were not consumptive, as well as on the effective probability that they fall ill. The heredity is affected by these probabilities, too.

Let us assume that at a certain stage a subject has a stock of knowledge allowing him to give certain answers with known probabilities. Before the next stage, the subject undergoes a learning process. Obviously, the probabilities of obtaining correct answers at this stage are not the same as at the preceding one, but are functions of these. Thus, chains with complete connections naturally arise in stochastic models for learning. A systematic account of the application of these models was given by Bush and Mosteller in their book [2].

When the chain x_n, $n \in N^*$ takes on a finite number of values $1, 2, \ldots, m$, the distribution function (14) has the form

$$F\left(x; \xi_{n-1}; \bar{p}_{\xi_1, \xi_2, \ldots, \xi_{n-2}}\right)$$

with

$$\bar{p}_{\xi_1, \xi_2, \ldots, \xi_{n-2}} = \left\{ p_{\xi_1, \xi_2, \ldots, \xi_{n-2}; 1}, \ldots, p_{\xi_1, \xi_2, \ldots, \xi_{n-2}; m} \right\},$$

where the $p_{\xi_1, \xi_2, \ldots, \xi_{n-2}; i}$ are the effective probabilities

$$p_{\xi_1, \xi_2, \ldots, \xi_{n-2}; i} = P(x_{n-1} = i | x_{n-2} = \xi_{n-2}, \ldots, x_1 = \xi_1)$$

of x_{n-1}. Thus we have a simple homogeneous finite state chain with complete connections if

$$p_{j_1, \ldots, j_{n-1}; j_n} = \varphi_{j_{n-1} j_n}\left(\bar{p}_{j_1, j_2, \ldots, j_{n-2}}\right)$$

whatever $1 \leqslant j_1, \ldots, j_n \leqslant m$, $n > 2$ and

$$p_{j_1; j_2} = \varphi_{j_1 j_2}(\bar{p})$$

where $\bar{p}$ is the probability vector of components

$$p_i = P(x_1 = i), \qquad 1 \leqslant i \leqslant m.$$

Here the functions φ_{ij}, $1 \leqslant i, j \leqslant m$ are defined on the simplex

$$D = \left\{ (p_i)_{1 \leqslant i \leqslant m} : 0 \leqslant p_i \leqslant 1, 1 \leqslant i \leqslant m, \sum_{i=1}^{m} p_i = 1 \right\}$$

and satisfy the conditions

$$0 \leqslant \varphi_{ij}(p_1,\ldots,p_m) \leqslant 1,$$

$$\sum_{j=1}^{m} \varphi_{ij}(p_1,\ldots,p_m) = 1,$$

whatever $i, j \in \{1,\ldots,m\}$.

With any chain with complete connections it is always possible to associate a Markov chain whose elements are now the sequence of random vectors $y_n = \bar{p}_{x_1,\ldots,x_n}$, $n \in N^*$. The relationship between a chain with complete connections and its associated Markov chain is underlined by the equation

$$P(x_n = k) = \sum_{j_1,\ldots,j_{n-1}} p_{j_1,\ldots,j_{n-1};k} = \sum_{j_1,\ldots,j_{n-1}} P_{j_1,\ldots,j_{n-1}} p_{j_1,\ldots,j_{n-1};k}$$

where

$$P_{j_1,\ldots,j_n} = P(x_1 = j_1,\ldots,x_n = j_n). \tag{16}$$

If, in learning theory, we refer to a class of answers $\{a_1,\ldots,a_m\}$, the probability that $x_n = a_k$ represents the probability of the event that the subject gives the answer a_k at the nth trial. The probability that $y_n = q_k$ represents the probability of the event that at the nth trial the subject gives the answer a_k with probability q_k.

By virtue of (16) the probability that $x_n = a_k$ is equal to the expectation of the kth component in the associated Markov chain. The probability that at the nth trial the subject gives the answer a_k equals the mean value of the probabilities with which he gives this answer, the number of these probabilities being m^{n-1}, i.e., the number of all the sequences of possible answers given at the previous $(n-1)$ trials.

In the case of a Markov chain, the influence exerted by the initial variable on the nth variable is diminished in most cases as n increases and is negligible for $n \to \infty$. The multiplicity of events influencing x_n leads to asymptotic regularity, known as the ergodic property. Professor Onicescu and I studied this property for chains with complete connections in the first paper devoted to them [27], giving sufficient conditions for ergodicity.

These conditions are as follows:

(i) The functions φ_{ij} are continuous on D, have continuous first derivatives and satisfy the conditions $0 < \varphi_{ij} < 1$, $1 \leqslant i, j \leqslant m$.

(ii) At least one of the transformations

$$(T_i) \quad p'_k = \varphi_{ik}(p_1,\ldots,p_m), \qquad 1 \leqslant k \leqslant m,$$

has a unique point of attraction in the interior of D.

(iii) The derivatives of the functions $\varphi_{ij}(p_1,\ldots,p_m)$ satisfy the inequalities

$$\sum_{j=1}^{m} \left| \frac{\partial \varphi_{ij}(p_1,\ldots,p_m)}{\partial p_k} \right| < 1$$

for any $1 \leqslant i, k \leqslant m$.

Under these conditions, denoting by $P_i^{(n)}(p_1,\ldots,p_m)$ the probability of the event $\{x_n = a_i\}$ where $p_i = P(x_1 = a_i)$, $1 \leqslant i \leqslant m$, we have

$$\lim_{n\to\infty} P_i^{(n)}(p_1, p_2,\ldots,p_m) = P_i, \qquad i = 1,2,\ldots,m.$$

The proof is based on the relation

$$P_i^{(n)}(p_1,\ldots,p_m) = \sum_k p_k P_i^{(n-1)}\left[\left(\varphi_{kj}(p_1,\ldots,p_m)\right)_{1\leqslant j\leqslant m}\right] \tag{17}$$

which extends the well-known Chapman–Kolmogorov equation for Markov chains to chains with complete connections.

Subsequently Fortet [3] gave weaker sufficient conditions for the ergodicity of chains with complete connections. He started from equation (17) written in the form

$$F_{n+1}(\bar{p}) = \sum_k p_k F_n[\varphi_k(\bar{p})], \tag{18}$$

and studied the asymptotic behaviour of $F_n(\bar{p})$ assuming $F_1(\bar{p})$ to be known. If the set D is endowed with a metric Δ, a transformation φ_k of D into itself is said to be bounded if there exists a constant $|\varphi_k|$ such that

$$\Delta[\varphi_k(\bar{p}'), \varphi_k(\bar{p}'')] \leqslant |\varphi_k|\Delta(\bar{p}', \bar{p}''), \qquad \bar{p}', \bar{p}'' \in D.$$

Fortet's conditions are that

(a) There exists a number L for which

$$|F_1(\bar{p}') - F_1(\bar{p}'')| \leqslant L\Delta(\bar{p}', \bar{p}''), \qquad \bar{p}', \bar{p}'' \in D;$$

(b) There exists a value k_0 for which

$$\varphi_{ik_0}(\bar{p}) > 0$$

for any i and $\bar{p} \in D$;

(c) The transformations φ_i are bounded and

$$|\varphi_i| < 1 \quad \text{for } i = k_0,$$
$$|\varphi_i| \leqslant 1 \quad \text{for } i \neq k_0.$$

Under these conditions the sequence of functions F_n converges to a constant F when n tends to ∞.

7. A School of Romanian Research Workers

The problem of ergodicity of chains with complete connections was also considered by other authors. I personally dealt with it in three papers [16], [17], [20] in which I examined the case of multiple chains and gave a Monte Carlo procedure for computing the limiting probabilities. The central limit theorem is also of interest for chains with complete connections: I considered it in [18] and [19].

At the end of the 1950s a very active period of research on dependence with complete connections began in Romania. In 1963, Marius Iosifescu considered a general framework for this type of stochastic dependence. In a paper distinguished by a Romanian Academy prize, he introduced the concept of a random system with complete connections. Such a system (in the homogeneous case) consists of a collection

$$((W,\mathcal{W}),(X,\mathcal{X}),u,P)$$

where $(W,\mathcal{W})$ and $(X,\mathcal{X})$ are measurable spaces, u is a mapping of $W\times X$ into W which is $(\mathcal{W}\times\mathcal{X},\mathcal{W})$-measurable, P is a transition probability function from $(W,\mathcal{W})$ to $(X,\mathcal{X})$ so that $P(w,\cdot)$ is a probability measure on $\mathcal{X}$ for any $w\in W$, and $P(\cdot,A)$ is a real-valued $\mathcal{W}$-measurable function for any $A\in\mathcal{X}$. With any random system with complete connections one can always associate two sequences of random variables $(x_n)_{n\geqslant 1}$ with values in X, and $(y_n)_{n\geqslant 0}$ with values in W such that $y_0=w_0$ (an arbitrarily fixed element of W), $y_{n+1}=u(y_n,x_{n+1})$, $n\geqslant 0$ and

$$P(x_1\in A)=P(w_0,A),\qquad A\in\mathcal{X},$$
$$P(x_{n+1}\in A\,|\,y_n,x_n,\ldots,x_1,y_0)=P(y_n,A),\qquad A\in\mathcal{X},\quad n\geqslant 1.$$

The sequence $(x_n)_{n\geqslant 1}$ is therefore an infinite-order chain while, as is easily seen, the sequence $(y_n)_{n\geqslant 0}$ is a Markov chain. The case of chains with complete connections is obtained by taking

$$X=\{1,2,\ldots,m\},$$
$$W=\left\{(p_j)_{1\leqslant j\leqslant m}: p_j\geqslant 0,\ \sum_{j=1}^{m} p_j=1\right\},$$
$$u\big((p_j)_{1\leqslant j\leqslant m};i\big)=\big(\varphi_{ij}(p_1,\ldots,p_m)\big)_{1\leqslant j\leqslant m},\qquad i\in X,$$
$$P\big((p_j)_{1\leqslant j\leqslant m};i\big)=p_i,\qquad i\in X.$$

We can see that multiple chains with complete connections, all stochastic models for learning, the continued fraction expansion, stochastic approximation procedures, and the Galton–Watson–Bienaymé processes with random environments can all be viewed as examples of random systems with complete connections.

Much of the work done up to 1969 by both Romanian and foreign probabilists on dependence with complete connections can be found in Iosifescu and Theodorescu's monograph [6]; Norman's book [26] should be considered as a useful complement to it. The work done since 1969 is included in a forthcoming book by Grigorescu and Iosifescu [5].

8. Actuarial Mathematics and Mathematical Statistics

I was especially interested in actuarial mathematics not only because of its current applications but also for its connection with other branches of mathematics. In my paper [13] I considered an insurance problem of a theoretical character. The insured person passes through the states $x \in R$, R denoting the real line, following a purely discontinuous stochastic process. The probability that an insured person in state x at time t will be in the same state at time $t + dt$ is

$$1 - p(x, t)\, dt + o(dt).$$

The individual may pass into another state with probability

$$p(x, t)\, dt + o(dt).$$

If an insured person is in the state $y \in R$ at time s $(s > t)$ he pays $\varphi(s, y)$; if at time s he passes from state y to state x, he receives $A(s, y, z)$.

From the equality of prospective and retrospective mathematical reserves, an equality which must be true whatever $\varphi(s, y)$ and $A(s, y, z)$, the Kolmogorov–Feller equations for purely discontinuous processes follow. I applied the same procedure to other more general stochastic processes occurring in insurance. In some demographic phenomena, the probability of the process being in state y at time s depends not only on the state x in which the process is at time $t (t < s)$, but also on the time it spent in this last state. For instance, when studying invalidity, the probability of an individual's getting out of the state of invalidity is a function both of this state and of the amount of time he has spent continuously in it. A new feature is thus introduced, making these processes a generalization of Markov processes. I studied this type of process in my joint paper with Firescu [24], extending the Kolmogorov equation to them.

I worked less in mathematical statistics: here I was mainly concerned with statistical inference for Markov processes and chains with complete connections [15], [19], [22], [23]. This does not mean that I do not appreciate mathematical statistics: quite the contrary. I contributed to the foundation of a research institute (the Centre of Mathematical Statistics of Bucharest) under the auspices of the Romanian Academy in 1964, and became its first director until 1976. A group of distinguished researchers today work in this institute, contributing effectively to the development of mathematical statistics and probability theory in Romania.

9. Concluding Remarks

Although I have already mentioned it, I wish to emphasize once more the point that I became involved in studying probability and its applications under the stimulus of Professor Onicescu. Had I not attended his 1928 lectures I would not have had the chance of capturing the flavour of the subject. This demonstrates the important rôle of first-rate mathematicians, who are not just makers of mathematics but also disseminators of mathematical ideas.

After a life devoted to science, I still feel I should have done more. This would have been possible if I could have spent less time in administrative and organizational work. On the other hand, I think that my involvement in such activities put me in a position where I was able to facilitate the access of deserving young people to research and teaching. Thus, at least in part, I do not regard these administrative duties as pure waste.

I am confident of the bright future of probability and statistics, and, more generally, of mathematics, in Romania. My expectations are based on the great number of gifted young people who work in almost all branches of mathematics. In the light of the ever-increasing importance of mathematics I am glad that Romania is in a good position to cope with the complexity of our present and future world.

Publications and References

[1] BERNSTEIN, S. (1926) Sur l'extension du théorème limite du calcul des probabilités aux sommes de quantités dépendantes. *Math. Ann.* **97**, 1–59.

[2] BUSH, R. and MOSTELLER, F. (1955) *Stochastic Models for Learning.* Wiley, New York.

[3] FORTET, R. (1938) Sur l'itération des substitutions algébriques linéaires à une infinité de variables et ses applications à la théorie des probabilités en chaîne. *Rev. Ci. (Lima)* **40**, 185–261, 337–447, 481–528.

[4] FRÉCHET, M. (1938) *Recherches théoriques sur le calcul des probabilités, Vol. II. Méthode des fonctions arbitraires. Théorie des événements en chaîne dans le cas d'un nombre fini d'états possibles.* Hermann, Paris.

[5] GRIGORESCU, S. and IOSIFESCU, M. (1982) *Dependence with Complete Connections and its Applications* (in Romanian).

[6] IOSIFESCU, M. and THEODORESCU, R. (1969) *Random Processes and Learning.* Springer-Verlag, Berlin.

[7] MARKOV, A. A. (1912) *Wahrscheinlichkeitsrechnung.* B. G. Teubner, Leipzig.

[8] MIHOC, G. (1935), (1936) On the general properties of dependent statistical variables (in Romanian) *Bull. Math. Soc. Roum. Sci.* **37** (1), 37–82; (2), 17–78.

[9] MIHOC, G. (1943) Sur le problème des itérations dans une suite d'épreuves. *Bull. Math. Soc. Roum. Sci.* **45**, 81–95.

[10] MIHOC, G. (1952) The law of rare events for Markov chains (in Romanian). *Bul. Şti. Acad. R.P.R. Sect. Şti. Mat. Fiz.* **4**, 783–790.

[11] MIHOC, G. (1954) Extending the Poisson law to homogeneous multiple Markov chains (in Romanian). *Bul. Şti. Acad. R.P.R. Sect. Sti. Mat. Fiz.* **6**, 5–15.

[12] MIHOC, G. (1956) Über verschiedene Ausdehnungen des Poissonschen Gesetzes auf endliche konstante Markoffsche Ketten. *Bericht Tagung Warhscheinlichkeitsrechnung Math. Statist. Berlin*, 43–49.

[13] MIHOC, G. (1956) An application of the theory of mathematical reserves in the theory of stochastic processes (in Romanian). *An. Univ. C. I. Parhon Ser. Şti.-Mat. Mat.-Fiz.* **12**, 13–18.

[14] MIHOC, G. (1956) Sur les lois limites des variables vectorielles enchaînées au sens de Markoff. *Teor. Verojatnost. i Primenen.* **1**, 103–112.

[15] MIHOC, G. (1957) Fonctions d'estimation efficiente pour les suites de variables dépendantes. *Bull. Math. Soc. Sci. Math. Phys. R. P. R.* **1** (49), 449–456.

[16] MIHOC, G. (1962) An ergodic theorem for multiple chains with complete connections (in Romanian). *An. Univ. Bucuresti Ser. Şti.-Nat. Mat.-Fiz.* **34**, 17–27.

[17] MIHOC, G. (1963) Sur l'application de la méthode de Monte-Carlo aux processus pour apprendre. *Rev. Roum. Math. Pures Appl.* **8**, 337–347.

[18] MIHOC, G. (1963) La loi limite pour les sommes des variables d'une chaîne à liaisons complètes, stationnaire et multiple. *Rev. Roum. Math. Pures Appl.* **8**, 217–226.

[19] MIHOC, G. (1976) Problèmes d'inférence statistique pour les chaînes à liaisons complètes. *Rev. Roum. Math. Pures Appl.* **21**, 1329–1332.

[20] MIHOC, G. and CIUCU, G. (1971) Sur l'ergodicité des chaînes à liaisons complètes. *Math. Balkanica* **1**, 164–170.

[21] MIHOC, G. and CIUCU, G. (1973) Sur la loi normale pour les chaînes à liaisons complètes. *Proc. Fourth Conf. Probability Theory Brasov, Romania*, Publishing House of the Romanian Academy, Bucharest, 169–171.

[22] MIHOC, G. and CRAIU, M. (1972) *Statistical Inference for Dependent Variables* (in Romanian). Publishing House of the Romanian Academy, Bucharest.

[23] MIHOC, G. and FIRESCU, D. (1959) Estimation functions for the parameters of a Markov process with transition density (in Romanian). *Ann. Univ. C. I. Parhon Ser. Şti.-Nat. Mat.-Fiz.* **22**, 9–16.

[24] MIHOC, G. and FIRESCU, D. (1962) A generalization of certain stochastic processes (in Romanian). *Comuničarile Acad. R. P. R.* **12**, 773–781.

[25] MIHOC, G. and NADEJDE, I. (1967) *Mathematical Programming*, Vol. II (in Romanian). Scientific Publishing House, Bucharest.

[26] NORMAN, M. F. (1972) *Markov Processes and Learning Models.* Academic Press, New York.

[27] ONICESCU, O. and MIHOC, G. (1935) Sur les chaînes de variables statistiques. *Bull. Sci. Math.* **59**, 174–192.

[28] ONICESCU, O. and MIHOC, G. (1940) Propriétés asymptotiques des chaînes de Markoff étudiées à l'aide de la fonction caractéristique. *Mathematica (Cluj)* **16**, 13–43.

[29] SCHULTZ, G. (1936) Grenzwertsätze für die Wahrscheinlichkeitenverketter Ereignisse. *Deutsche Math.* **1**, 665–699.

2
STATISTICIANS IN STOCHASTIC PROCESSES AND INDEPENDENCE

M. S. Bartlett

Maurice Stevenson Bartlett was born in London in 1910 and received his secondary education at Latymer Upper School. After reading mathematics at Queens' College, Cambridge he began his career in 1933 as Assistant Lecturer in the Department of Statistics at University College London. He joined ICI in 1934 as a statistician and worked there for four years before being appointed to a lectureship in mathematics at Cambridge in 1938. During the Second World War, he was engaged on rocket research; shortly after its end, in 1947, he was appointed to the Chair of Mathematical Statistics at the University of Manchester, where he built up a lively department. In 1960, he moved to the Professorship of Statistics at University College London and in 1967 became Professor of Biomathematics at Oxford; he retired from this post in 1975.

Professor Bartlett has worked with distinction in many areas of statistics and stochastic processes. He is the author of five books and numerous research papers and was elected to Fellowship of the Royal Society in 1961. Among his many honours are the Guy Medal in Gold of the Royal Statistical Society (1969), the University of Oxford's Weldon Medal (1971), and honorary D.Sc.s from both the University of Chicago (1966) and the University of Hull (1976).

He has been most influential in forming the British school of stochastic analysts, and has trained many Ph.D. students throughout the world. He has been President of the Manchester Statistical Society (1959–60) and of the Royal Statistical Society (1966–67), and has recently been created Honorary Member of the International Statistical Institute (1980).

He was married to Sheila Chapman in 1957; they have one daughter. His hobbies are landscape gardening and collecting water-colours.

Chance and Change

M. S. Bartlett

1. Cambridge Days

At the beginning of my book on *Stochastic Processes* [23] is a brief quotation from Shelley's *Prometheus Unbound*; hence my even briefer extract as my present title, which perhaps epitomizes my career, both personally and professionally.

My interest in probability began at school with the chapter in Hall and Knight's *Algebra*, and developed further at Cambridge when my third-year courses for Part III of the Mathematical Tripos included not only Wishart's "Statistics" course (which I took the first year it was given), but also Eddington's "Combination of Observations," and Fowler's "Statistical Mechanics" (where probabilities were introduced most discreetly as "weights"). At that time, as my undergraduate days at Cambridge drew to an end, I was anxious to be finished with study, and what I felt to be a surfeit of mathematics, and to get a job; I elected to try for the Home Civil Service (administrative grade), with the Inland Revenue as a second string. Statisticians were not recognised as such in the Civil Service until later, and although statistics was a subject in the HCS examination it was, with its 100 marks maximum, insufficient to make up my required quota, and I was obliged to take physics instead. I did well in the Inland Revenue examination, but was not good enough in the HCS to qualify for a vacancy (no mathematicians were that year, a change of examiners making things more difficult for us!). This was, however, my first stroke of luck, for my state scholarship was renewed for a fourth year; I decided not to take up the Inland Revenue post, but to return to Cambridge with the view of trying again for the HCS.

This enabled me to become Wishart's first mathematical postgraduate student, continuing with the study, and even research [74], in mathematical statistics begun in my final undergraduate year. I attended further lecture courses for "fun," including Eddington's "Relativity" and Dirac's "Quantum Mechanics," as well as Colin Clark's "Statistical Sources" (which I found rather dull) and Udny Yule's "Vital Statistics" (still given in spite of his retirement). At the Union Library I discovered the book by F. P. Ramsey [69] on the foundations of mathematics, which included succinct remarks on the nature of probability, chance and likelihood. Ramsey had hardly started as my lecturer in mathematical analysis in my first undergraduate year, when a fatal illness cut short first his lectures, and then his life, when he was still a young man. He was the only lecturer in analysis I listened to at Cambridge who seemed able to combine rigour with simplicity; and, unfortunately for my pure mathematics, I subsequently found the turgid style of some others, perhaps inevitable with the current syllabus, which was based on Riemannian integration, rather off-putting. (G. H. Hardy was a Cambridge professor then, but I did not hear him lecture, as professors did not normally lecture for Parts I and II of the Tripos.)

Towards the end of my postgraduate year I was beginning to think of a career in statistics as much more attractive than an administrative post in the Civil Service, even if I secured one; so that when E. S. Pearson, with Wishart's recommendation, approached me about a vacancy in his new Statistics Department at University College, London, I was quite ready to accept, beginning in October 1933.

2. University College and Imperial Chemical Industries

This was not to say that I found my first job, at University College, ideal. I had already met R. A. Fisher during a "pilgrimage" to Rothamsted (see [41]) and his attitude to statistics, as something to be *used*, reflected also in Wishart's link with the Cambridge School of Agriculture, appealed to me strongly. In fact, mathematical theory only made sense to me at that time if it was to be applied, an echo no doubt of the fine applied mathematics tradition I had absorbed from Cambridge, even if my pure mathematics was more restricted; my undergraduate tutor at Queens' College, Cambridge, had been A. Munro, very much an "applied" man.

At University College I was fortunate in the people around me, as I recalled later in my inaugural lecture on my return there in 1960 (see [38], p. 3). R. A. Fisher had just arrived as the new Galton Professor. J. Neyman had joined E. S. Pearson in his department, and I was able to hear his lectures on the new Neyman–Pearson theory of testing statistical hypotheses; and J. B. S. Haldane was frequently about in Fisher's department. Nevertheless, my task of "preaching" statistics before having adequately learned to "practise" made me feel uneasy, especially as teaching as such has never been all that attractive to me. I was not particularly worried about the prickly relation between the Statistics Department and the Galton Laboratory upstairs. This had resulted in a splitting of the tea-interval, which took place in the Common Room on the first floor, into two contiguous intervals for the two departments; but by judicious timing one was able to overlap and talk to members of the other department. I had just written what I still regard as one of my best papers [1], on the vector structure of analysis of variance (and its multivariate extension), deriving almost incidentally the correct test for covariance adjustment in an experimental design and analysis. Fisher's existing discussion in the fourth edition of *Statistical Methods for Research Workers* [49] was rather misleading on this point, and when I showed him my draft I was a little disappointed to be told firmly that a revision was already in hand for the next edition. However, I was able to mention this in a footnote!

What I did not realize with the multivariate generalization in this paper was the possibility of a *canonical* transformation of variables, this being first introduced by Hotelling [58] in relation to *correlations*, though I subsequently [9] emphasized its value in terms of the wider regression approach.

My wish to get to grips more directly with statistical methods was granted after only one year at University College, when I was offered the post of statistician at the Agricultural Research Station at Jealott's Hill run by Imperial Chemical Industries Ltd. This Research Institute, while concerned largely with ICI's interests in artificial fertilizers, was, when I joined it, organized somewhat on Rothamsted lines, with links with overseas experimentation as well as with its local research. The director was imbued with the broad scientific Rothamsted tradition, and I personally was able to absorb something of the agricultural and biological background at Jealott's Hill while maintaining my theoretical interest in the role of statistics and probability in science at large. There were no constraints on my external correspondence, which included contact with J. B. S. Haldane, following up some work with him on the theory of inbreeding (see [56], [57]), Godfrey H. Thomson, and even R. H. Fowler and A. S. Eddington, from my Cambridge interest in statistical physics. This led to some problems in the filing department, as it was a laudable custom in ICI to register centrally each letter, with a reply, or at least an acknowledgment, expected within two days. It also meant that external correspondence was their property, and was retained when I later left Jealott's Hill, though an exception was made with one letter from Eddington, which I still have with the ICI filing stamp on it!

Retrospectively, I would say that my time at Jealott's Hill was not only the happiest period (professionally) of my life, but also my most creative. I am sure that my own work was rather undisciplined through the previous lack of any systematic research training for a higher degree with an official supervisor; nevertheless, the research background seemed to be a great mental stimulus not only to research on directly related topics, but to much else as well. I even published a letter in *Nature* [3] on the concept of time; and while this may have manifested more than most my lack of research discipline it did precede the work of E. A. Milne [64] on the arbitrariness of the time-scale and the effect of transformations of it. (On looking through Milne's paper I see that he gives a reference to de Sitter [70], p. 185.) It elicited two written responses, the one I have mentioned from Eddington, and one from H. T. Pledge, author of *Science Since 1500* [67].

More orthodox contributions included my examination of sufficiency properties in the case of more than one parameter [4], [6], and my controversy with Fisher arising out of this on his extension of fiducial theory to the multiparameter case. I first tried to argue about this in correspondence (which Graham Wilkinson intends, I understand, to include in Fisher correspondence to be published), believing that his apparent conditioning on an estimated variance s^2 was mistaken, as s^2 is *not* sufficient for σ^2 (cf. my fuller remarks in [31]). Only after I had failed to convince him did I decide to publish [5] and much later was astonished to hear from Wishart that Fisher resigned from the Cambridge Philosophical Society when they published my paper.

My general admiration for Fisher's work began to be tempered with the suspicion that he did not like to admit mistakes, although the intuitive and heuristic type of mathematics which he favoured (and with which I was much in sympathy) does tend to be more prone to error. Quite apart from the controversy about fiducial probability, which remained unresolved because of differing interpretations, there were one or two other exchanges on problems which were less debatable. These included:

(a) contingency table interactions [2]: here Fisher was helpful when I wrote to him concerning an inadequate analysis by T. N. Hoblyn, of the East Malling Research Station, so that I did not emphasise, and Fisher did not mention, that he had suggested the previous analysis;

(b) multivariate analysis [9], [10]: when I sent Fisher a copy of my 1938 paper referring to approximate tests in canonical analysis, he sent me a typescript note, correcting the erroneous degrees of freedom given in a paper of his own [50]; this note he claimed had inadvertently been overlooked in the printing, but in any case Fisher's comments were still misleading in implying that the test was exact. In the same paper he also gave a wrong test for a hypothetical discriminant function, and was rather grudging in his comment on my correction [10]: "...This is true, but confusing..." (Fisher [51], p. 423).

It was a calamity for me when a sudden decision by ICI's chief research scientist in London drastically reduced their entire research budget throughout the country and thereby initiated in particular the departure of many of the staff at Jealott's Hill, including our director. I was also "axed," though not in so extreme a manner, being put on the transfer list. I landed up in the Intelligence Department at Head Office, where my job was to co-ordinate the production returns for the whole of ICI as required by the Board of Trade. I had no strong objection to my new job, at least for the time being; it was perhaps regarded as promotion, but my new environment put an effective damper on research. So when after a year in London I was approached about a vacancy at Ames with G. Snedecor, I considered it seriously and sought permission from ICI for leave of absence to see how I liked it. However, as this permission was not granted, I decided to apply instead for a vacant lectureship in mathematics at Cambridge. This incidentally released the Ames vacancy for W. G. Cochran, who in consequence emigrated to the United States.

3. Cambridge Again, and the War Years

The Cambridge post was especially slanted to statistics, and I was appointed to it in October 1938; but I found that I was required to give some other lecture courses as well, in mechanics and in elementary mathematical analysis. After some years away from academic life, I found preparing all these lectures hard work. To make matters worse, I belatedly discovered that the topic for the Adams Prize Mathematical Essay, which is set in Cambridge every two years, and was due for submission by the end of the year, was one in statistical theory. It was foolhardy of me to contemplate trying to cope with it as well, all in my first term, especially as the title set: "The distributional properties of functions of statistical variables" was not too clear about its intended scope. In spite of my daily schedule of work having to be curtailed to allow for exercise, on medical advice following signs of mental strain, I produced an entry of sorts, but was hardly surprised when it proved unsuccessful. A rather dubious consolation was that no one was successful (I do not know what other entries, if any, there were). I did receive, however, a sympathetic letter from the adjudicators, A. C. Aitken, A. S. Eddington and E. C. Whittaker, urging me to publish my entry. In fact I never did this, for shortly afterwards M. G. Kendall approached J. O. Irwin, E. S. Pearson, J. Wishart and myself about co-operating in writing a treatise on theoretical statistics (cf. my obituary article on E. S. Pearson [42]). This co-operative project was in turn abandoned when war broke out in 1939, and the treatise completed eventually by Kendall on his own.

University staff were allowed to be allocated to war jobs on a temporary basis, retaining their tenure and right to reinstatement; and it was in the hard winter at the end of 1939 or the beginning of 1940 that I set off for a Government establishment in Kent engaged on research and development in rocketry. A considerable part of my time was spent in assessing the effectiveness of rocket weapons for various defensive and offensive purposes, and my particular statistical expertise was very relevant, both in analysing trials and in theoretical calculations, though all of the work was naturally concerned with immediate and urgent applied research and development.

My work base was shifted at frequent intervals. After the initial period at Fort Halstead, I was moved with other mathematicians to a house in Sevenoaks, where for a while I shared a room with D. R. Hartree. On the fall of France we were all evacuated to Wales, but some of us then came back to form a headquarters department in London, just in time for the beginning of the bombing "blitz!" Later some of us moved back yet again to Wales. At headquarters I had acquired F. J. Anscombe (a research student of mine at Cambridge) as an assistant at one stage; but back in Wales D. G. Kendall, who was one of the wartime group of mathematicians there, became my assistant and consequently began to learn statistics. Another mathematician in the group was P. A. P. Moran, but he did not become converted to statistics until the war was over.

It was during the war years that I also first met J. E. Moyal, through our mutual interests rather than by chance encounter. I had, as part of my general interest in the role of probabilistic ideas in statistical physics, always been puzzled by the anomalous way in which probability had slipped into the new wave mechanics, not fundamentally but as an interpretation of the positive measure $\psi\psi^*$, where ψ is the wave function. I heard, I think through J. O. Irwin (who was in Cambridge at the time), that Moyal, who had previously been in France, had been working on this problem; and this was to be the start of a long association between us. Progress with the wave or quantum-mechanical problem was slow and limited (cf. my article "The paradox of probability in physics" in [38]), but Moyal's more general knowledge of European work in the theory of stochastic processes was a considerable stimulus to me. As I have remarked before, English statisticians tended for a long time to believe that a traditional empiricism exonerated them from overmuch study of abstract continental mathematics. Unfortunately in the case of stochastic processes this resulted in a big vacuum in our education, and in particular my own tentative efforts with time series (stimulated by Yule's work) and other temporal models before the war would have been greatly aided by earlier knowledge of Kolmogorov's fundamental work [61], and also of Khintchine's 1934 paper [60] on stationary processes. (Regarding the latter, cf. my remarks in [40]; of course the universal ignorance and neglect of McKendrick's work on biological processes, e.g. [63], even in England, was an additional obstacle to any comprehensive and rapid progress.)

I find I still have an extensive collection of letters from Jo Moyal around 1943–4, when he was stationed in Leicester, concerning his work. While it would not be appropriate here to refer to this correspondence in detail, it is relevant to recall that he had a fairly protracted discussion at my instigation with R. H. Fowler, and also some with P. A. M. Dirac and H. Jeffreys, about the best way to publish. The result seems adequately summarized by the following passage from a letter to me dated 27 June 1944:

> "...I do not think I told you about my meeting with Fowler. What he finally suggested was that the parts of my paper not dealing with quantum theory should appear in the form of a book. Professor Hardy and the Cambridge Press have now accepted it for publication as a monograph of about 200 pages. It will, of course, have to be extended to include an account of all the important work on the subject. I have not yet started doing anything about it, and if you want to change your mind about my suggestion of collaboration, I shall be very pleased..."

4. After the War at Cambridge

When the war was over, it was a little while before it was possible to return to Cambridge; but when I got back in 1946 I was very much immersed in work on time series and stochastic processes. In the first lecture course I gave in this area I discussed two main topics (a) time series cf. [12]; (b) diffusion processes, including reference to their use in asymptotic sequential sampling theory [11], which was no longer classified material!

It was at this time that Harold Hotelling moved from Columbia to Chapel Hill, North Carolina, to lead the theoretical side of the two-department Institute of Statistics, the applied department being at Raleigh. I was invited to join Hotelling's team. With my customary caution I first sought leave of absence for a year from Cambridge to go to Chapel Hill. Technically, I thought I had qualified for this leave, as I had been nominally on the Cambridge "establishment" throughout the war years, but it was perhaps not surprising that so long a leave shortly after my return was not granted; however, I was granted four months, and travelled across the Atlantic in the *Queen Mary* in company with a shipful of Canadian war brides, enjoying, with some risk to the digestion, my first spell of unrationed food since the war had initiated food rationing.

In Chapel Hill I opted to lecture again on stochastic processes, (see [13]), combining this time population processes with the previous topic of time series. It was in these lectures that I included the solution of the simple birth-and-death process in continuous time introduced by Feller [48], using an "operational" approach for the probability-generation function that was familiar to me in the case of previous discrete-time problems (see Appendix 1, p. 55). When I returned to Cambridge at the beginning of 1947 (having rejected the idea of permanent transfer to Chapel Hill), I generalized this approach for Markov processes in general, and wrote it up at Moyal's suggestion with a view to publication. However, it was rejected, primarily as Palm [66] had already used similar methods to solve the birth-and-death process, but partly also on charges such as "incomprehensibility," an accusation which I naturally resented (my reply to the referee, and a relevant letter to D. G. Kendall, are perhaps of some historic interest, and are reproduced in Appendix 2).

This general characteristic function version of the "forward" equation was later noted in my contribution [17] to the 1949 Royal Statistical Society Symposium on Stochastic Processes, organized in co-operation with D. G. Kendall and J. E. Moyal. It was somewhat later that the alternative use made of the backward equation by L. Jànossy [59] in the study of cosmic-ray showers was extended by David Kendall and myself to study the characteristic functional of what we termed "regenerative" processes [44]; we noted in our paper that such processes had also been introduced by Palm [66].

The wartime co-operation with Jo Moyal had culminated in a mutual decision to collaborate in writing a general book on stochastic processes, a decision apparently not taken until 1946 (see preface to [23]). When this project dragged on year after year, and other books on stochastic processes began to appear, we agreed, however, on a separation of effort, and my book, *Stochastic Processes* [23], finally appeared in print in 1955.

The immediate post-war years were obviously a busy professional period after the frustration of the war years; and, while not perhaps having the youthful exuberance of the Jealott's Hill era, my research papers were no doubt (apart from the ill-fated 1947 paper) more mature in content. They included not only my paper on time series [12], and my 1949 paper [17] for the Stochastic Processes Symposium, but also my multivariate analysis paper on the general canonical correlation distribution [14]. This paper I regarded as important in going a long way to solving a difficult distributional problem, rather than in the practical value of the results, the problem having existed since the introduction of canonical correlations by Hotelling. The geometrical approach I used was an extension of Fisher's approach to obtain the distribution of the multiple correlation coefficient distribution, an approach which, as I noted in my Fisher Memorial Lecture [31], I had failed, in company with others, to understand before the war, although I succeeded in my later post-war attempt.

This search for the solution almost as an intellectual exercise was somewhat remote from my undergraduate philosophy of finding only useful mathematics interesting; but even Fisher, in spite of his avowed aversion to purely "academic" results, was quite ready at times for such indulgences. It seems at least more excusable than the pride displayed by some mathematicians in never soiling their hands with "practical mathematics," in contrast with some of the most illustrious, for example, Kolmogorov, Turing, Norbert Wiener and von Neumann, who have contributed to both theoretical and applied mathematics.

5. Manchester

Before the end of the academic year 1946–7, I was approached about a new Chair of Mathematical Statistics created at Manchester University, a post I felt honoured to occupy. Manchester had a reputation in science, and the two mathematics professors, Max Newman and Sidney Goldstein, had both moved from Cambridge with the aim of organizing a department worthy of Manchester's tradition. The North of England was strange territory for me, but I found that the people seemed to compensate for any deficiencies in the climate, and showed a warm-heartedness not always so manifest in the south. On the academic side there were initial discussions on how to include statistics in the undergraduate curriculum, especially as I was anxious for practical work to be included in the syllabus. The problem was resolved by getting statistics recognized as one of two subsidiary subjects selected in the second year, with the option of further specialization within the main mathematics curriculum in the third year. A later innovation at the post-graduate level was the Statistics Diploma course, based on the same objectives as the one I had had a hand in formulating at Cambridge. This arose at Manchester after a discussion with P. M. S. Blackett, then Professor of Physics there, on the possibility of introducing a postgraduate course on operational research. This idea I was rather against, in spite of the proven value of the subject during the war, arguing that in a university it was more reasonable to inculcate basic disciplines such as statistics or economics as a prelude to future careers in management or operational research. In due course university courses were started elsewhere in operational research; but the difficulty some of them have had in maintaining a genuine and balanced applied content, as distinct from being a hotch-potch of separate courses on statistics, information theory, stochastic processes, linear programming, and so on, has at least not dispelled my doubts about the feasibility of an integrated university curriculum.

My period of thirteen years at Manchester from 1947 enabled me to consolidate further my study and research in time series and stochastic processes, already begun after the war in Cambridge and North Carolina. Among fields of application, I had already indicated in the 1949 RSS Symposium the relevance to epidemic theory (being still ignorant at that time of McKendrick's earlier work in this field), and with the help of my very able computing assistant, Mrs L. Linnert, collected various statistics on measles incidence in England and Wales, including figures for Manchester wards and boroughs from the local records. I was particularly interested in possible mechanisms for maintaining recurrent epidemics in diseases like measles, and this work, first reported at Berkeley [24], which was a Mecca for mathematical statisticians during the years of Jerzy Neyman's famous symposia, culminated in my papers on measles periodicity in the *Journal of the Royal Statistical Society* [26], [28]. A feature of interest, both theoretically and practically, was the discovery of a "critical community size," below which local extinction of infection was a dominating feature of the epidemic pattern.

J. E. Moyal had joined me on the statistics staff at Manchester, and among our research students Alladi Ramakrishnan, who had worked with H. J. Bhabha on the theory of cosmic-ray showers, came over from India to work on the theory of point processes. These processes had been introduced in the context of time series by Herman Wold, and in effect by David Kendall, in connection with the characteristic functional for the age variable in his paper on population processes for the 1949 RSS Symposium; but it was Ramakrishnan who emphasized the special rules to be employed in handling quantities like $dN(t) = N(t + dt) - N(t)$, as distinct from a continuous process $X(t)$, introducing the notion of "product densities."

The publication of my book on *Stochastic Processes* in 1955 must have stimulated the invitation to the Soviet Mathematical Congress in 1956 (Mary Cartwright being the other mathematician invited from England), for I discovered (unofficially) after my arrival that it was being produced in translated form in Russia. I have reported on this visit elsewhere [25]. The opportunity of meeting A. N. Kolmogorov, G. F. Gause (author of *The Struggle for Existence*) and others (though not A. Khintchine) was certainly welcome to me, though I have never been sure how far my own paper on stochastic models in biology was particularly welcome at the conference at that particular time.

In 1958 I was able to get a term's leave of absence to visit Harvard, where W. G. Cochran had joined F. Mosteller. This was a pleasant interlude which must have enabled me to complete the little monograph on stochastic population models [27], and also to obtain some North American measles statistics, with the help of Jane Worcester, in order to confirm the epidemic pattern I had previously reported for England and Wales.

Research in other areas of statistics continued steadily. The post-war impetus starting with the 1946 RSS Symposium had led in due course to my paper on smoothing periodograms ([18], reported more briefly in a letter to *Nature* [16]), and, following earlier work by U. Grenander, to methods of working with the *cumulative* periodogram [22]. Grenander had also, in a fundamental paper in 1950 [55], raised the problem of inference for stochastic processes, dealing especially with the problems arising from continuous-time processes. In a different region of the same general area, I discussed the inference problem for Markov chains, including "higher-order" chains [19]. One feature of this work was the natural way one could extend the use of the likelihood function familiar from more classical problems.

Another area still of interest was multivariate analysis; and when E. J. Williams proposed the use of some exact tests in the context of hypothetical canonical variates, I showed by geometrical arguments that this class of solution was generally available [21], in contrast with the approximate tests available making use of *sample* canonical variates. A. M. Kshirsagar, who was a research student with me at Manchester, later confirmed these results by analytical methods, and simplified some of the computational formulae (see, for example, [62]).

In spite of the attraction of these exact methods, there are some problems and dangers in the elimination of *hypothetical* variates. The danger is that tests of residual variation are not valid unless the eliminated variate is the *true* one (cf. [38], p. 146); and a problem not altogether resolved, and much wider than this particular context, is the apparent gain in degrees of freedom compared with the situation when *sample* variates are eliminated, a gain which seems to become more suspect if sample variates are used to control the class of hypothetical variates permitted.

Other visitors to Manchester included J. Gani, as a Nuffield Research Fellow from Australia (cf. his preface to [53]).

6. University College Again

In 1960 Egon Pearson retired from the Statistics Chair at University College, London, and I was invited to succeed him. In spite of the division of the original chair endowed by Galton, and of the creation of new statistics professorships in London, for example, at the London School of Economics, Birkbeck, and Imperial College, there was still a historical lustre about the University College post which persuaded me to accept it, although my time at Manchester had been very satisfying.

The London organization of colleges proved in fact to be rather daunting. University College was still the only college offering a full-time degree in statistics, but administratively and educationally, the University of London had its combined councils and committees, in addition to those of each institution, and this implied a somewhat heavy load of committees for departmental heads.

Nevertheless, research in time series and stochastic models continued, perhaps more sedately. At least the university vacations were largely free of commitments; and a country cottage in Suffolk, acquired after moving from Buxton to an apartment in Highgate, led to an efficient division of my vacation days between work in the morning and gardening in the afternoon. It was during one such vacation that work was begun on my RSS paper [29], on the extension of spectral analysis to point processes, an extension which, if used with suitable care (for example, correlation or spectral analysis is not such an "exhaustive" analysis as it is for Gaussian processes), has proved a useful technique, later generalized the two-dimensional and multivariate processes [30], [36].

In 1964 I was invited by Professor Paul Meier to visit Chicago for a term; it was during this period that I was sounded about joining the Statistics Department there, but in spite of a tempting offer my links with England were too strong to be broken permanently.

7. Oxford

Nevertheless, after seven years the prospect of continuing until retirement in London was not all that exciting; and when a new Chair of Biomathematics was created at Oxford, with the responsibility of developing biological statistics and mathematics, the post seemed enough of a challenge for me to apply. It was perhaps rash of me, for Oxford had not shown any great generosity to their statisticians in the past, or much rational organisation of statistics. The existence of an Institute of Statistics, even though it was orientated to economic statistics, had precluded the use of the word "statistics" for posts elsewhere in the university, so that D. J. Finney had been appointed to a "Lectureship in the Design and Analysis of Scientific

Experiment," and, when he left, N. T. J. Bailey to a "Readership in Biometry," before he in turn moved elsewhere.

Links with P. H. Leslie, who was with the Bureau of Animal Population at Oxford, and with G. E. Blackman, the Professor of Agriculture, and an erstwhile colleague many years before at Jealott's Hill, were promising credit items (another contact of interest was Dr R. E. Hope-Simpson, who lived in Cirencester); but on the debit side I soon discovered that the Biomathematics Department and its staff occupied a rather uneasy position in the general teaching structure, and were viewed suspiciously by some biologists as being too mathematical. The mathematicians seemed more generally ready to welcome me, but I had moved to Oxford to try to foster biological statistics and mathematics and I knew that this aim would not be served by drifting even further away from the biological side.

The "struggle for existence" became a real one as the financial climate worsened, even the traditional advisory functions of the staff being threatened by the "staff–student ratio" calculations popular as a prelude to quinquennial budget and staffing recommendations by the university.

On the research side, my general aim, as I indicated in my inaugural lecture (see [38]), was to cover theoretical *population* biology, together with biometry on the statistical side, these areas being naturally associated. This overall area has, I am glad to say, stabilized as the recognized area of interest of the department.

As regards personal research, my efforts seemed to become channelled almost more than I wanted into a study of two-dimensional spatial systems, culminating eventually into a small monograph on spatial analysis [39]. This interest had already been referred to in an Appendix to my Presidential Address to the Royal Statistical Society [32], where I started examining the status and values of "one-sided" probability chains in two dimensions. (The situation for continuous variable systems had already been discussed by P. Whittle [72] as long before as 1954.) This had a dual purpose (a) in seeking simple stochastic models for biometrical spatial data (b) in hoping for a possible simpler approach to multidimensional physical lattice systems, the mathematics for which was notoriously difficult (Onsager [65]).

The processes I looked at in 1967 did not look too promising for (b), not even being spatially symmetrical; and, a few years later, stimulated by some lectures in Oxford by Mark Kac, I investigated classes of spatial-temporal models which did look more promising, in that even "one-sided" spatial interaction could lead to the required equilibrium distribution. Unfortunately, they were, perhaps not surprisingly, non-linear processes, as I reported at the 1970 Mathematical Congress [33] in Nice (see also [34], [35], [37]), and did not facilitate any exact solution, though approximate solutions were possible by this approach in the three-dimensional as well as the two-dimensional case. This was rather disappointing, in view of the known exact Onsager solution in the two-dimensional case. More recently "one-sided" spatial chains have been further studied by others, two-dimensional symmetric versions being possible, and even three-dimensional versions *leading to a phase transition* (see, for example Welberry and Miller [71]). In the study of these models, Julian Besag was with me at Oxford as a research assistant for a while, and discovered some important properties in relation to their status as Markov fields (see Besag [45], [46]).

In 1973–4 I qualified for a whole sabbatical year away from Oxford and went, at Professor Moran's invitation, to his department in Canberra, Australia. This was the one and only whole year of "sabbatical" leave I have ever had, and was especially welcome both to myself and my wife, in her case because of exhaustion after nursing her mother (who suffered a stroke about 1970), and in mine because of a hope of writing up my study of spatial processes as a monograph. My own aim was somewhat handicapped by hospitalization in Canberra for prostate trouble, and an operation which went wrong, leading to indifferent health and three further operations in the years following my return to England. This influenced my decision to retire in 1975, two years earlier than might otherwise have been the case. Nevertheless, a return to better health has enabled me to emerge into a somewhat more active period again, including two enjoyable return visits to Australia, one in 1977 and one in 1980. This last visit was particularly valuable in giving me an opportunity to catch up a bit with neglected reading, for example, in theoretical population biology; I was also pleased to link some work on "doubly stochastic" population models with my much earlier work on birth-and-death processes and the like (see Appendix to Bartlett [43]).

8. Concluding Remarks

Looking back over developments of interest to me, I would, I suppose, have to classify an increasing number under stochastic process theory and modelling rather than under statistics as such. Conscious at times of limitations in my pure mathematical ability, and handicapped to start with by ignorance of some of the relevant literature, I have nevertheless always felt the need for a much more comprehensive body of theory fusing probability with processes in time, in most branches of science. At least, this overall long-term view has been amply justified by developments; the days, for instance, when "stochastic processes" was an unknown phrase to Cambridge mathematicians have long since gone.

The situation for mathematical statisticians has in consequence become less clear, for the problems of matching models with data have become more complicated and technical, and less straightforward. Statistical inference for me has always been an open-ended problem, which does not lend itself easily to "tidy" answers (Bayesian or otherwise). Yet the applied tradition of matching observation with theory or model is one that none of us should forget.

APPENDIX 1

Comments on the Use of Generating Functions in Stochastic Process Theory

The use of generating functions in mathematics and statistics goes back to the time of Laplace and Lagrange, but was particularly evident in H. W. Watson's discussion of the "extinction of surnames" problem (see Appendix to Galton [52]). In the theory of inbreeding which was at the time treated deterministically (e.g. [56], [57]) I was concerned in [7] to have a fully *stochastic* version of population changes from generation to generation; I summarize the following opening passage:

Let the chance of an individual of type Aa arising out of the self-fertilization of a heterozygous plant be $\frac{1}{2}$. Let the generating function of the statistical moments of the probability distribution of the number n_s of Aa in the sth generation be $M_s(t)=E(e^{tn_s})$, where E denotes expectation. If the ith plant gives rise to f_i offspring,

$$M_{s+1}=E\prod_i\left(\tfrac{1}{2}+\tfrac{1}{2}e^t\right)^{f_i}. \tag{1}$$

Let the generating function of f_i be the function $F(t)$. Then by definition

$$E\left(\tfrac{1}{2}+\tfrac{1}{2}e^t\right)^{f_i}=F(\theta), \tag{2}$$

where $\theta=\log(\frac{1}{2}+\frac{1}{2}e^t)$, and

$$M_{s+1}=E\prod_i F\left(\log\left[\tfrac{1}{2}+\tfrac{1}{2}e^t\right]\right). \tag{3}$$

This has to be averaged for variation in the number n_s of factors in (3). Since any function $EG(n_s)$ may be written symbolically

$$\left[G\left(\frac{\partial}{\partial\tau}\right)M_s(\tau)\right]_{\tau=0}, \tag{4}$$

we obtain finally

$$M_{s+1}(t)=\left\{F\left(\log\left[\tfrac{1}{2}+\tfrac{1}{2}e^t\right]\right)\right\}^{\partial/\partial\tau}M_s(\tau)_{\tau=0}, \tag{5}$$

τ being put equal to 0 after the differentiation.

In the two special cases (a) of a fixed number f of offspring, so that $F=e^{ft}$ (b) f is a Poisson variable with mean g, so that $\log F=g(e^t-1)$, equation (5) is equivalent to

$$M_{s+1}(t)=M_s(\phi),$$

where

$$\text{(a) } \phi=f\log\left(\tfrac{1}{2}+\tfrac{1}{2}e^t\right),$$

$$\text{(b) } \phi=\tfrac{1}{2}g(e^t-1).$$

A simpler stochastic process example in discrete time is discussed in Bartlett [9]. From a number n of coins on a table, a random coin is turned over, and the process repeated continually. Then, if r_s is the number of 'heads' showing after s reversals, and $M_s(t)=E(e^{tr_s})$, we at once obtain

$$M_{s+1}(t)=E\left(\frac{r_s}{n}e^{t(r_s-1)}+\frac{n-r_s}{n}e^{t(r_s+1)}\right)$$
$$=\left\{e^t\left(1-\frac{1}{n}\frac{\partial}{\partial t}\right)+\frac{1}{n}e^{-t}\frac{\partial}{\partial t}\right\}M_s,$$

whence

$$M_s(t)=\left\{e^t\left(1-\frac{1}{n}\frac{\partial}{\partial t}\right)+\frac{1}{n}e^{-t}\frac{\partial}{\partial t}\right\}^s M_0.$$

In equilibrium $M_{s+1}(t)=M_s(t)$, so that

$$\frac{1}{n}(1+e^{-t})\frac{\partial M}{\partial t}=M,$$

whence

$$M=\left(\tfrac{1}{2}+\tfrac{1}{2}e^t\right)^n.$$

I first used similar methods for *continuous-time* processes in [13], cf. also [15], [17], reference being made in the latter to the prior work of C. Palm [66]. (My reply to a referee of Bartlett [15], and a letter to David Kendall, are reproduced in Appendix 2 below.) E. Feldheim [47] also used the characteristic function for stochastic processes, but only for diffusion processes, investigating the conditions leading to normal distributions (conditions later discovered independently by Whittle [73]).

APPENDIX 2

Correspondence Relating to Bartlett [15]

137, Chesterton Rd,
Cambridge
Aug 21, '47

Dear David,

I have just had (this evening) an unpleasant shock in the rejection of my *Annals* paper. The discovery that the differential equation solution of the λ, μ process has apparently already been solved (by Palm) and published, (the referee refers to Arley & Borschenius, *Acta Math* 76 p. 299, which I have yet to look at), is enough to knock the bottom out of my paper, but I rather resent the tone of the report, which speaks of a "large mass of undigested & undigestible material quite unsuitable for publication" with "no new results".

I thought I would let you know my paper is K.O.'d at once, as you will require to modify your own papers (I believe your *Annals* one will be O.K.) accordingly.

It is possible that Feller is the referee, as the latter also refers to the solution being given by Feller in a lecture in January, 1946, and if so maybe (with perhaps some justification) he feels I did not take sufficient trouble to check up on his and Arley's recent work; but why on earth he or Arley didn't tell me this when I raised the subject at Princeton I don't know.

Yrs
Maurice Bartlett

Some Comments on the Referee's Report on My Paper "The Rate of Change of the Characteristic Function in the Theory of Stochastic Processes"

The referee's recommendation of rejection is accepted in view of the prior solutions of the λ, μ process. The tone of his report seems to me, however, astonishingly severe for a paper written in good faith, and his accusations of "incomprehensibility" at least flatly contradicted, as far as I am aware, by other readers of the typescript.

The two main points in the paper, as indicated in the summary, were:

(i) the demonstration of the use of the partial differential equation method for the characteristic function in solving the λ, μ process. I must apologize for not being aware (in common with many others) of its previous solution. Arley (1943) and Feller (1939) had failed to reach a solution. I mentioned the generating function technique to them in conversation at Princeton in November, 1946, but unfortunately for me they did not, while of course saying that this was a useful method, make it clear to me that this method had already been made use of in this context and an account of it published.

(ii) the setting-up of the general equation for the characteristic function $\phi(\theta; t) \equiv E\{e^{i\theta X(t)}\}$ viz:

$$\frac{\partial \phi(\theta; t)}{\partial t} = \int e^{i\theta x} \psi(i\theta, t, x)\, dF(x; t)$$

when

$$\psi(i\theta, t, x) = \operatorname*{Lt}_{\Delta t \to 0} \int \frac{(e^{i\theta y} - 1)\, dH(y = \Delta x(t) | x; \Delta t, t)}{\Delta t}$$

exists.

There is perhaps nothing very subtle in setting-up this straightforward relation, but neither I nor Moyal (who has a considerable experience of the related literature) were aware of its previous discussion. Moyal has made an important application of this result in theoretical physics, and it was in fact at his suggestion that my paper was 'written-up' with a view to publication.

The use already made of my paper by D. G. Kendall, Moyal and others seems to me a sufficient refutation of some of the general charges levelled at it by the referee. The question of priority is another matter, and I would like to thank the referee for his explicit references, which were new to me.

M. S. Bartlett
21/8/47

Publications and References

[1] BARTLETT, M. S. (1934) The vector representation of a sample. *Proc. Camb. Phil. Soc.* **30**, 327–340.

[2] BARTLETT, M. S. (1935) Contingency table interactions. *J. R. Statist. Soc. Suppl.* **2**, 248–252.

[3] BARTLETT, M. S. (1936a) Intrinsic uncertainty of reference frames. *Nature London* **138**, 401.

[4] BARTLETT, M. S. (1936b) Statistical information and properties of sufficiency. *Proc. R. Soc. London* A **154**, 124–137.

[5] BARTLETT, M. S. (1936c) The information available in small samples. *Proc. Camb. Phil. Soc.* **32**, 560–566.

[6] BARTLETT, M. S. (1937a) Properties of sufficiency and statistical tests. *Proc. R. Soc. London* A **160**, 268–282.

[7] BARTLETT, M. S. (1937b) Deviations from expected frequencies in the theory of inbreeding. *J. Genetics* **35**, 83–87.

[8] BARTLETT, M. S. (1938a) The distributional properties of functions of statistical variables. Adams Prize entry, unpublished.

[9] BARTLETT, M. S. (1938b) Further aspects of the theory of multiple regression. *Proc. Camb. Phil. Soc.* **34**, 33–40.

[10] BARTLETT, M. S. (1939) A note on tests of significance in multivariate analysis. *Proc. Camb. Phil. Soc.* **35**, 180–185.

[11] BARTLETT, M. S. (1946a) The large sample theory of sequential tests. *Proc. Camb. Phil. Soc.* **42**, 239–244.

[12] BARTLETT, M. S. (1946b) On the theoretical specification and sampling properties of autocorrelated time series. *J. R. Statist. Soc. Suppl.* **8**, 27–41.

[13] BARTLETT, M. S. (1946c) Stochastic Processes. Mimeographed notes of a course given at the University of North Carolina.

[14] BARTLETT, M. S. (1947a) The general canonical correlation distribution. *Ann. Math. Statist.* **18**, 1–17.

[15] BARTLETT, M. S. (1947b) The rate of change of the characteristic function in the theory of stochastic processes. Unpublished.

[16] BARTLETT, M. S. (1948) Smoothing periodograms from time-series with continuous spectra. *Nature, London* **161**, 686.

[17] BARTLETT, M. S. (1949) Some evolutionary stochastic processes. *J. R. Statist. Soc.* B **11**, 211–229.

[18] BARTLETT, M. S. (1950a) Periodogram analysis and continuous spectra. *Biometrika* **37**, 1–16.

[19] BARTLETT, M. S. (1950b) The frequency goodness of fit test for probability chains. *Proc. Camb. Phil. Soc.* **47**, 86–95.

[20] BARTLETT, M. S. and KENDALL, D. G. (1951) On the use of the characteristic functions in the analysis of some stochastic processes occurring in physics and biology. *Proc. Camb. Phil. Soc.* **47**, 65–76.

[21] BARTLETT, M. S. (1951) The goodness of fit of a single hypothetical discriminant function in the case of several groups. *Ann. Eugen.* **16**, 199–214.

[22] BARTLETT, M. S. (1954) Problèmes de l'analyse spectrale des series temporelles stationnaires. *Publ. Inst. Statist. Univ. Paris* **3**, 119–134.

[23] BARTLETT, M. S. (1955) *An Introduction to Stochastic Processes.* Cambridge University Press.

[24] BARTLETT, M. S. (1956a) Deterministic and stochastic models for recurrent epidemics. *Proc. 3rd Berkeley Symp. Math. Statist. Prob.* **4**, 81–109.

[25] BARTLETT, M. S. (1956b) Note on a visit to Moscow for the Third Soviet Mathematical Congress. *J. R. Statist. Soc.* A **119**, 456.

[26] BARTLETT, M. S. (1957) Measles periodicity and community size. *J. R. Statist. Soc.* A **120**, 48–70.

[27] BARTLETT, M. S. (1960a) *Stochastic Population Models in Ecology and Epidemiology.* Methuen, London.

[28] BARTLETT, M. S. (1960b) The critical community size for measles in the United States. *J. R. Statist. Soc.* A **123**, 37–44.

[29] BARTLETT, M. S. (1963) The spectral analysis of point processes. *J. R. Statist. Soc.* B **25**, 264–296.

[30] BARTLETT, M. S. (1964) The spectral analysis of two-dimensional point processes. *Biometrika* **51**, 299–311.

[31] BARTLETT, M. S. (1965) R. A. Fisher and the last fifty years of statistical methodology. *J. Amer. Statist. Assoc.* **60**, 395–409.

[32] BARTLETT, M. S. (1967) Inference and stochastic processes. *J. R. Statist. Soc.* A **130**, 457–477.

[33] BARTLETT, M. S. (1970) Physical nearest-neighbour models and non-linear time-series. Internat. Congr. Mathematicians (Nice).

[34] BARTLETT, M. S. (1971) Physical nearest-neighbour models and non-linear time series. *J. Appl. Prob.* **8**, 222–232.

[35] BARTLETT M. S. (1972a) Physical nearest-neighbour models and non-linear time series II. Further discussion of approximate solutions and exact equations. *J. Appl. Prob.* **9**, 76–86.

[36] BARTLETT, M. S. (1972b) Some applications of multivariate point processes. In *Stochastic Point Processes*, ed P. A. W. Lewis, Wiley, New York, 136–150.

[37] BARTLETT, M. S. (1974) Physical nearest-neighbour models and non-linear time series III. Non-zero means and sub-critical temperatures. *J. Appl. Prob.* **11**, 715–725.

[38] BARTLETT, M. S. (1975) *Probability, Statistics and Time.* Chapman and Hall, London.

[39] BARTLETT, M. S. (1976) *The Statistical Analysis of Spatial Pattern.* Chapman and Hall, London.

[40] BARTLETT, M. S. (1978) Correlation or spectral analysis? *The Statistician* **27**, 147–158.

[41] BARTLETT, M. S. (1980) All our yesterdays. Newsletter 69, CSIRO Division of Mathematics and Statistics.

[42] BARTLETT, M. S. (1981a) Egon Sharpe Pearson, 1895–1980. *Biometrika* **68**, 1–11.

[43] BARTLETT, M. S. (1981b) Population and community structure and interaction. *Proc. Conf. Mathematical Theory of the Dynamics of Biological Populations.* Academic Press.

[44] BARTLETT, M. S. and KENDALL, D. G. (1951) On the use of characteristic functions in the analysis of some stochastic processes occurring in physics and biology. *Proc. Camb. Phil. Soc.* **47**, 65–76.

[45] BESAG, J. E. (1972) Nearest neighbour systems and the autologistic model for binary data. *J. R. Statist. Soc.* B **34**, 75–83.

[46] BESAG, J. E. (1974) Spatial interaction and the statistical analysis of lattice systems. (with discussion) *J. R. Statist. Soc.* B **36**, 192–236.

[47] FELDHEIM, E. (1936) Sur les probabilités en chaîne. *Math. Ann.* **112**, 775–780.

[48] FELLER, W. (1939) Die Grundlagen der Volterraschen Theorie des Kampfes ums Dasein Wahrscheinlichkeitstheoretischer Behandlung. *Acta Biotheoretica* **5**, 11–40.

[49] FISHER, R. A. (1932) *Statistical Methods for Research Workers*, 4th edn. Oliver and Boyd, Edinburgh.

[50] FISHER, R. A. (1938) The statistical utilization of multiple measurements. *Ann. Eugen.* **8**, 376–386.

[51] FISHER, R. A. (1940) The precision of discriminant functions. *Ann. Eugen.* **10**, 422–429.

[52] GALTON, F. (1889) *Natural Inheritance*. Macmillan, London.

[53] GANI, J., ED. (1975) *Perspectives in Probability and Statistics*. Distributed by Academic Press, London, for the Applied Probability Trust, Sheffield.

[54] GAUSE, G. F. (1934) *The Struggle for Existence*. Baltimore.

[55] GRENANDER, U. (1950) Stochastic processes and statistical inference. *Ark. Mat.* **1**, 195–277.

[56] HALDANE, J. B. S. and BARTLETT, M. S. (1934) The theory of inbreeding in autotetraploids. *J. Genetics* **29**, 175–180.

[57] HALDANE, J. B. S. and BARTLETT, M. S. (1935) The theory of inbreeding with forced heterozygosis. *J. Genetics* **31**, 327–340.

[58] HOTELLING, H. (1936) Relations between two sets of variates. *Biometrika* **28**, 321–377.

[59] JÀNOSSY, L. (1950) Note on the fluctuation problem of cascades. *Proc. Phys. Soc. London* A **63**, 241–249.

[60] KHINTCHINE, A. (1934) Korrelationstheorie der stationaren stochastische Prozesse. *Math. Ann.* **109**, 604–615.

[61] KOLMOGOROV, A. N. (1931) Über die analytische Methoden in der Wahrscheinlichkeitsrechnung. *Math. Ann.* **104**, 415–458.

[62] KSHIRSAGAR, A. M. (1972) *Multivariate Analysis*. Dekker, New York.

[63] MCKENDRICK, A. G. (1926) Applications of mathematics to medical problems. *Proc. Edinburgh Math. Soc.* **44**, 98–130.

[64] MILNE, E. A. (1937) Kinematics, dynamics and the scale of time. *Proc. R. Soc. London* A **158**, 324–348.

[65] ONSAGER, L. (1944) Crystal statistics I. A two-dimensional model with an order-disorder transition. *Phys. Rev.* **63**, 117–149.

[66] PALM, C. (1943) Intensitätsschwangungen im Fernsprechverkehr. *Ericsson Technics* **44**, 1–189.

[67] PLEDGE, H. T. (1939) *Science Since 1500*. HMSO, London.

[68] RAMAKRISHNAN, A. (1950) Stochastic processes relating to particles distributed in a continuous infinity of states. *Proc. Camb. Phil. Soc.* **46**, 595–602.

[69] RAMSEY, F. P. (1931) *The Foundations of Mathematics, and Other Logical Essays*. Routledge and Kegan Paul, London.

[70] SITTER, W. DE (1933) *The Astronomical Aspect of the Theory of Relativity*. University of California Press, Berkeley.

[71] WELBERRY, T. R. and MILLER, G. H. (1978) A phase transition in a 3D growth-disorder model. *Acta Cryst.* A **34**, 120–123.

[72] WHITTLE, P. (1954) On stationary processes in the plane. *Biometrika* **41**, 434–449.

[73] WHITTLE, P. (1957) On the use of the normal approximation in the treatment of stochastic processes. *J. R. Statist. Soc.* B **19**, 268–281.

[74] WISHART, J. and BARTLETT, M. S. (1932) Distribution of second-order moment statistics in a normal system. *Proc. Camb. Phil. Soc.* **28**, 455–459.

Mark Kac

Mark Kac was born in Krzemieniec, Poland in 1914 and became fascinated at an early age by Euclidean geometry in which he was tutored by his father. He studied mathematics and physics at the John Casimir University in Lwów and obtained his Ph.D. in 1937.

After graduating, he worked as an actuary at the Phoenix insurance company, and in 1938 went to the USA to do research at The Johns Hopkins University. In 1939, he joined the mathematics faculty of Cornell University and rose to become full Professor in 1947. In 1961 he joined the Rockefeller University and in 1981 he moved to the University of Southern California where he is currently professor.

Professor Kac has exceptionally wide interests in probability and its relationship to the physical sciences, as well as to other branches of mathematics. He is noted for dramatizing mathematics and its relation to the rest of the world, and has been awarded two Chauvenet Prizes by the Mathematical Association of America for his expository papers on "Random Walk and the Theory of Brownian Motion" and "Can One Hear the Shape of a Drum?"

Professor Kac has been active in mathematical and statistical affairs both in the USA and internationally. He has been a member of the Council of the American Mathematical Society and Editor of its *Transactions* between 1955 and 1958. He was President of the Institute of Mathematical Statistics in 1981.

Among his many honours are election to the American Academy of Arts and Sciences in 1959 and the National Academy of Sciences in 1965. He received an honorary doctorate from the Case Institute of Technology in 1966, and was elected foreign member of the Royal Norwegian Academy of Science and member of the American Philosophical Society in 1969.

Professor Kac has travelled widely in his capacity as visiting professor: to Leiden, Oxford, Brussels, Utrecht and Pisa, among other universities. He is married, and has a son and a daughter. His recreations are bridge and golf, both of which he plays badly.

The Search for the Meaning of Independence

Mark Kac

To the memory of Hugo Steinhaus

A few years ago, while the publication of a selection of my papers was in preparation, a young colleague who was invited by the editors to write a commentary on some of the papers came to see me. He was in some distress because much of an early paper appeared trivial to him, and yet he was reluctant to make a negative comment on a youthful contribution of one of his teachers. He compromised by not commenting at all, and it was left to me in an autobiographical sketch which I wrote for the selection to make a few remarks on "Sur les fonctions indépendantes I" which was the paper in question.

The paper appeared in 1936 in *Studia Mathematica* although most of the material was worked out in the summer of 1935. In August of that year I also reached my twenty-first birthday.

At the time, probability theory was not taught and, in fact, not even considered to be a bona fide part of mathematics. The appearance in 1933 of the book by Kolmogorov on foundations of probability theory was hardly noticed in Lwów, where I was a student. I do remember looking at the book, but I was frightened away by its abstractness, and it did not seem to have much to do with what I had been working on. I was wrong, of course, but about this later.

What I was concerned with was the study of real-valued measurable functions $f_1(t), f_2(t), f_3(t), \ldots$ defined on the interval $(0,1)$ and satisfying the condition that for all real $\alpha_1, \alpha_2, \alpha_3, \ldots$ and for every integer n

$$\mu\{f_1(t)<\alpha_1, f_2(t)<\alpha_2, \ldots, f_n(t)<\alpha_n\} = \prod_{k=1}^{n} \mu\{f_k(t)<\alpha_k\} \tag{1}$$

with $\mu\{\cdot\}$ denoting the Lebesgue measure of the set defined inside the braces. Steinhaus proposed that I study these independent functions, as he called them, and almost at once I began to rediscover much that was already known in an abstract setting to Kolmogorov, Khintchine, Paul Lévy, Doob and a few (very few) others.

We now teach young students that random variables are measurable functions defined on a measure space, and independence of random variables is embodied in (1) with t ranging over that measure space. In 1936, however, I thought that I made a discovery of some importance when I wrote in the final sentence of my paper: "Il est assez clair que l'on peut identifier les 'variables éventuelles indépendantes', avec les fonctions indépendantes; cette interprétation nous permettra de réduire les notions du calcul des probabilités à celles de la théorie des fonctions des variables réelles." "Variables éventuelles indépendantes," i.e. independent random variables were to me (and others, including my teacher Steinhaus) shadowy and not really well-defined objects which were now, for the first time, made part of a well-established theory of functions of a real variable.

This view may (and does) appear strange from the vantage point of today, but it was perfectly natural if one knows some of the background.

In 1922 Steinhaus published a fundamental memoir [11], in which the first rigorous theory of the game of heads or tails was given. In modern terms, Steinhaus proved the existence of a product measure in the product space $A \times A \times \cdots$, where A consisted of two points $(0,1)$ (or (H,T)) with each point assigned measure $\frac{1}{2}$. This he accomplished by mapping the space $A \times A \times \cdots$ (i.e. the space of all sequences $\varepsilon_1, \varepsilon_2, \cdots$ of zeros and ones) onto the interval $0 \leqslant t \leqslant 1$ by the formula

$$t = \frac{\varepsilon_1}{2} + \frac{\varepsilon_2}{2^2} + \ldots, \tag{2}$$

and by showing that the product measure maps thereby on the ordinary Lebesgue measure in the interval $0 \leqslant t \leqslant 1$.

Steinhaus was motivated by the question "What is the probability that the series

$$\sum_{n=1}^{\infty} \pm c_n \qquad (c_n \text{ real numbers}) \tag{3}$$

with 'random' signs converges?" "Random" was to be interpreted as being determined by independent tosses of an "honest" coin. Rewriting (2) in the more illuminating form

$$t = \frac{\varepsilon_1(t)}{2} + \frac{\varepsilon_2(t)}{2^2} + \cdots, \tag{4}$$

and setting

$$r_n(t) = 1 - 2\varepsilon_n(t), \tag{5}$$

Steinhaus noted that his question concerning the series (3) could now be translated into asking for the measure of the set of t's for which the series

$$\sum_{n=1}^{\infty} c_n r_n(t) \tag{6}$$

converges. In this form the question was already (partially) answered by Rademacher (in the same year 1922), who proved that the measure in question is 1 provided that

$$\sum_{n=1}^{\infty} c_n^2 < \infty. \tag{7}$$

The title of Rademacher's paper [10], in which he introduced the functions $r_n(t)$ now justly bearing his name, was "Einige Sätze über Reihen von allgemeinen Orthogonalfunktionen." It was thus in the context of orthogonal series, a subject of considerable interest during the 1930s and 1940s, that the Rademacher functions entered the mathematical literature. Rademacher once told me in a casual conversation that he strongly suspected that his functions had something to do with the game of "heads or tails," but he was dissuaded from further pursuing the subject by Richard von Mises who, as we know, had his own way of looking at chance and probability.

Steinhaus did recognize in 1922 the probabilistic significance of Rademacher functions, but even he did not appreciate then that the Rademacher functions are *independent* (in the sense of (1)). If I may be permitted a repetition of a mild witticism (used as the heading of Section 1.4 of Chapter I of my Carus Monograph *Independence in Probability, Analysis and the Theory of Numbers*) Steinhaus failed to recognize that $1/2^n$ is $1/2$ multiplied by itself n times. Had Steinhaus realized that the Rademacher functions are independent he would have had no difficulty proving the counterpart of Rademacher's theorem, i.e. that (6) diverges almost everywhere provided that

$$\sum_{n=1}^{\infty} c_n^2 = \infty. \tag{8}$$

For assuming that (6) converges on a set of positive measure it would have to converge almost everywhere (by the 0–1 principle which, in a weaker form, was already known) and thus denoting by $g(t)$ the sum of (6) we would have for every real $\xi \neq 0$

$$\int_0^1 \exp(i\xi g(t))\, dt = \int_0^1 \exp\left(i\xi \sum_{n=1}^{\infty} c_n r_n(t) \right) dt = \prod_{n=1}^{\infty} \cos c_n \xi. \tag{9}$$

Under the additional condition $c_n \to 0$ (otherwise the almost everywhere divergence would be trivial) the infinite product in (9) is equal to 0 and therefore

$$\int_0^1 \exp(i\xi g(t))\, dt = 0 \tag{10}$$

for all real $\xi \neq 0$, and this is easily seen to be impossible.

But then characteristic functions like

$$\int_0^1 \exp\left(i\xi \sum_{n=1}^{\infty} c_n r_n(t) \right) dt \tag{11}$$

were outside the lore of orthogonal series, and even if (11) had somehow been written down, it is unlikely that without full appreciation of what independence implies, it would have been noticed that it is equal to the product

$$\prod_{n=1}^{\infty} \cos c_n \xi.$$

As it is, the counterpart of Rademacher's theorem had to wait for a proof until 1925, when Kolmogorov and Khintchine [8] provided one (in greater generality). The Kolmogorov–Khintchine paper is quoted in the definitive monograph *Theorie der Orthogonalreihen* by Kaczmarz and Steinhaus [5], but the proof given there is based on an elegant idea of Zygmund. This, though applicable also to the proof that (8) implies divergence almost everywhere of

$$\sum_{n=1}^{\infty} c_n \sin 2\pi m_n t \tag{12}$$

provided the sequence $\{m_n\}$ has Hadamard gaps, i.e.

$$\frac{m_{n+1}}{m_n} > q > 1, \tag{13}$$

unfortunately obscures the role of independence.

Even more surprising is a 1930 paper by Steinhaus [12] in which he introduced the functions $\vartheta_n(t), 0 \leqslant t \leqslant 1$, which are independent in the sense of (1), and uniformly distributed in (0, 1), i.e.

$$\mu\{a < \vartheta_n(t) < b\} = b - a$$

for all a, b such that $0 \leqslant a < b \leqslant 1$. He then used these to prove that the series

$$\sum_{n=1}^{\infty} c_n \exp(2\pi i \vartheta_n(t))$$

converges for almost all t provided

$$\sum_{n=1}^{\infty} |c_n|^2 < \infty.$$

The same functions have also been introduced by Wiener who also shares with Steinhaus the idea of considering series $\Sigma \pm c_n$ with $+$, $-$ signs assigned at random.

This is a very special case of the so-called three series theorem of Kolmogorov [6], [7]. The title of Kolmogorov's paper was "Über die Summen durch den Zufall bestimmter unabhängiger Grössen," and Steinhaus either overlooked it or, more likely, did not realize that his $\exp(2\pi i \vartheta_n(t))$ were but an example of "unabhängiger Grössen." In my paper I merely followed Steinhaus, step by step, and proved that for any square-integrable, real-valued, independent functions $f_n(t)$ such that

$$\int_0^1 f_n(t)\,dt = 0, \qquad n = 1, 2, \ldots, \tag{14}$$

the series

$$\sum_{n=1}^{\infty} f_n(t)$$

converges almost everywhere provided

$$\sum_{n=1}^{\infty} \int_0^1 f_n^2(t)\,dt < \infty.$$

This too was a special case of Kolmogorov's theorem, but the role of independence was at last recognized (by me, that is).

That my thinking (like that of Steinhaus) was influenced by the problematics of the theory of orthogonal series is clearly seen by my devoting a number of pages in my note to the "wrong way" Schwarz inequality. This inequality is the following: If again $f_n(t)$ are independent (real-valued) square-integrable functions such that in addition to (14) they satisfy the inequality

$$b_n^2 = \int_0^1 f_n^2(t)\,dt > \frac{1}{q}\int_0^1 f_n^4(t)\,dt, \tag{15}$$

then

$$\int_0^1 \left| \sum_{k=1}^{n} f_k(t) \right| dt > R(q) \left(\sum_{k=1}^{n} b_k^2 \right)^{1/2}, \tag{16}$$

with the constant $R(q)$ depending only on q. Schwarz's inequality gives, of course,

$$\int_0^1 \left| \sum_{k=1}^{n} f_k(t) \right| dt \leqslant \left(\sum_{k=1}^{n} b_k^2 \right)^{1/2} \tag{17}$$

Now, one may ask, why (16)? It has no probabilistic interest whatsoever, but it happens to be one of the criteria of "lacunarity" of an orthonormal system (see e.g. [5]), and for Rademacher functions as well as for the functions $\{\sin 2\pi m_n t\} m_{n+1}/m_n > q > 1$, it was known.

The most striking display of ignorance on my part is Théorème 2. This theorem states that if again $f_n(t)$ are independent satisfying (14), and if

$$b_n^2 = \int_0^1 f_n^2(t)\,dt = 1, \qquad n = 1, 2, \ldots$$

and

$$\int_0^1 |f_n^k(t)|\,dt < M^k, \qquad k = 2, 4, \ldots$$

(in the original paper M^k is misprinted as M), then

$$\lim_{n\to\infty}\int_0^1\left(\frac{f_1(t)+\cdots+f_n(t)}{\sqrt{n}}\right)^k dt=\frac{1}{\sqrt{2\pi}}\int_{-\infty}^{\infty}u^k\exp(-u^2/2)\,du. \quad (18)$$

The inclusion of this well-known result is surprising on two counts: first I did not draw from (18) the immediate conclusion that

$$\lim_{n\to\infty}\mu\left\{\frac{\sum_{k=1} f_k(t)}{\sqrt{n}}<\omega\right\}=\frac{1}{\sqrt{2\pi}}\int_{-\infty}^{\omega}\exp(-u^2/2)\,du, \quad (19)$$

i.e. a very weak form of the central limit theorem and, second, I did not refer to Markov's book *Wahrscheinlichkeitsrechnung*, which I had been reading at the time, and by which I was truly fascinated. My Théorème 2 was taken straight from Markov, but I did not fully appreciate that (18) implies (19) and it had not yet completely dawned on me that the—to me—mysterious "independent quantities" X_n of Markov were modelled by the wonderfully concrete independent functions $f_n(t)$.

People to whom something wonderful happens sometimes pinch themselves in search of reassurance that they are not dreaming. The realization that the mysterious

$$\frac{1}{\sqrt{2\pi}}\int_{-\infty}^{\omega}\exp(-u^2/2)\,du \quad (20)$$

might soon become part of ordinary analysis hit me with such a force that I "pinched myself" (in print!), by verifying that I was not dreaming and that independent functions rooted safely in the theory of orthogonal systems were providing a bridge to the exotic and elusive world of randomness and chance.

The attitude of mathematicians in the 1930s to the "law of errors" (20) was not much different from that wittily expressed by Poincaré in 1912: "Tout le monde y croit (la loi des erreurs) cependant, me disait un jour M. Lipmann, car les expérimentateurs s'imaginent que c'est un théorème de Mathématiques, et les mathématiciens, que c'est un fait expérimental." Suddenly we had it in our hands to cast a decisive vote with the experimentalists. The "loi des erreurs" that everybody believed in could after all turn out to be "un théorème de Mathématiques."

But, the reader could justly ask, was the central limit theorem not already known? It was, of course, but not in the sense that it really "belonged." It was either a somewhat esoteric asymptotic result on binomial coefficients (de Moivre, Laplace) or a theorem on convolutions (Lindeberg, Feller) or, finally a statement about not-well-defined "zufälliger Grössen" which did not become clearer by being translated into French as "variables aléatoires" (Markov, Liapunov, Lévy). Even when random variables became rigorously defined as measurable functions on a measure space, and independence with equal rigor associated with product measures, the central limit theorem remained in the dark outer edges of mathematics, neither widely understood nor much appreciated. Contrary to the common belief, the association of independence with product measures is due to Z. Łomnicki and S. Ulam. Although their paper [9] appeared in 1934, one year after the appearance of Kolmogorov's book, the principal results were presented at the 1932 International Congress of Mathematicians in Zürich.

This, Steinhaus and I thought, was now going to change, and after once again "pinching ourselves" in print by rediscovering a somewhat less special case of the central limit theorem than my Théorème 2, as well as Paul Lévy's method of proof based on characteristic functions, we finally produced an analytic result of unquestioned novelty and interest.

In our note [4], submitted in October 1936, only nine months after my own note was submitted, we introduced the natural concept of *relative measure* μ_R of a set E on the real line as

$$\mu_R(E) = \lim_{T\to\infty} \frac{\mu\{(-T,T)\cap E\}}{2T}, \tag{21}$$

provided the limit exists.

We then proved that if $\lambda_1, \lambda_2, \ldots$ are linearly independent (independent over the field of rationals), then for example, the functions $\cos\lambda_k t$ are independent in the sense of (1) with μ being replaced by μ_R. Now our "pinching ourselves" bore fruit, for it followed almost at once that

$$\lim_{n\to\infty} \mu_R\left\{\sqrt{2}\,\frac{\sum_{k=1}^{n}\cos\lambda_k t}{\sqrt{n}} < \omega\right\} = \frac{1}{\sqrt{2\pi}}\int_{-\infty}^{\omega} \exp(-u^2/2)\,du. \tag{22}$$

Lest the modern reader fails to be impressed by (22), let me state a slight extension of this result.

Setting

$$x_n(t) = \sqrt{2}\,\frac{\sum_{k=1}^{n}\cos\lambda_k t}{\sqrt{n}} \tag{23}$$

and assuming the linearly independent λ_k are such that

$$\lim_{n\to\infty} \frac{k_n(\lambda)}{n} = A(\lambda), \tag{24}$$

where $k_n(\lambda)$ is the number of λ_k's among the first n which are less than λ, we have that the joint distribution of

$$x_n(t+\tau_1), x_n(t+\tau_2), \ldots, x_n(t+\tau_n)$$

in the limit as $n \to \infty$ is multivariate Gaussian with the covariance matrix $((\rho(\tau_i - \tau_j)))$ $(i, j = 1, 2, \ldots, n)$, where

$$\rho(\tau) = \int_0^\infty \cos \lambda \tau \, dA(\lambda). \tag{25}$$

In other words, as $n \to \infty$, $x_n(t)$ becomes *indistinguishable* from a *typical* sample path of a stationary Gaussian process $x(t)$ with the (integrated) power spectrum $A(\lambda)$. And yet there is nothing "random" in the definition of $x_n(t)$ which is as "deterministic" as anything could possibly be. We thus see that (22) as extended achieves a measure of philosophical significance, since it casts serious doubt on the premise glibly accepted by philosophers and experimentalists alike, that there may be some operationally discernible distinction between random and deterministic processes.

Philosophical considerations aside, (22) was right in the center of mathematics, since independence of $\cos \lambda_k t$ is also an easy consequence of the classical Kronecker–Weyl theorem; it should be noted that cos can be replaced by a much more general function and still have independence, but for the validity of (25) cos is essential. In this discovery we were anticipated by A. Wintner [13], who however did not draw the immediate conclusions (22). It is also interesting to note that Wintner in the title of his note speaks of statistical independence of *distribution functions* of the $\cos \lambda_k t$, not of the cosines themselves, another indication, albeit a small one, of how unclear the concept of independence was in the 1930s. As a purely autobiographical comment, I should mention that it was the partial confluence of our ideas with those of Wintner and his collaborators that made me choose the Johns Hopkins University at which to hold the fellowship which I was awarded in 1938 by the Parnas Foundation of Lwów, Poland.

Once it became clear that independence is a property of functions which need not have to live in some abstract (sample) spaces but could be just ordinary functions defined on such prosaic spaces as the interval $(0, 1)$ or the infinite line, a new chapter in analysis began to unfold.

I have already mentioned the application of the central limit theorem to $\cos \lambda_k t$ with linearly independent frequencies. Shortly after, in [3], I proved that if $\phi(t), 0 \leqslant t \leqslant 1$; satisfies a Hölder condition and is normalized so that

$$\int_0^1 \phi(t)\, dt = 0, \int_0^1 \phi^2(t)\, dt = 1, \tag{26}$$

then

$$\lim_{n \to \infty} \mu \left\{ \frac{\sum_{k=1}^{n} \phi(2^{k(k-1)/2} t)}{\sqrt{n}} < \omega \right\} = \frac{1}{\sqrt{2\pi}} \int_{-\infty}^{\omega} \exp(-u^2/2)\, du. \tag{27}$$

Theorems of this kind would have been unthinkable only a few years earlier—and it is interesting to note that after more than forty years, their reverberations are still felt.

I have particularly in mind the recent preoccupation with phenomena exhibited by iterations of relatively simple functions, and in particular with the most spectacular of these phenomena which goes by the name of "chaos." "Chaos" is just a word suggested by the sometimes observed wildly irregular appearance of the sequence of iterates.

In at least one case there is more to "chaos" than picturesqueness and this is the case of the iterates of the function f defined on $(0,1)$ as follows:

$$f(t)=2t, 0\leqslant t\leqslant\tfrac{1}{2}, f(t)=2(1-t), \tfrac{1}{2}\leqslant t\leqslant 1. \tag{28}$$

Setting

$$f^{(1)}(t)=f(t), f^{(k+1)}(t)=f(f^{(k)}(t))$$

one has

$$\lim_{n\to\infty}\mu\left\{\sqrt{12}\,\frac{\sum_{k=1}^{n}\left(f^{(k)}(t)-1/2\right)}{\sqrt{n}}<\omega\right\}=\frac{1}{\sqrt{2\pi}}\int_{-\infty}^{\omega}\exp(-u^2/2)\,du, \tag{29}$$

a central limit theorem for iterates!

In the special case (28) it happens that

$$f^{(k)}(t)\equiv f(2^k t) \tag{30}$$

(if one extends f periodically with period 1), and it also happens that my theorem (27) is also valid with $k(k+1)/2$ replaced by k. This was first discovered by Fortet [2] in 1940 under the influence of my theorem, and there is now an extensive literature on this subject.

The most fruitful results of having understood the notion of independence and its implications are, however, to be found in number theory. I briefly reviewed part of the story at a recent conference held at the North Texas University, Denton, Texas, to celebrate the famed "Scottish Book" of problems.[1] The English translation of the "Scottish Book," together with commentaries and a number of papers based on talks delivered at the conference (including my own), will soon be published by Birkhauser Verlag. Let me quote here the final few paragraphs:

"I cannot remember at all how I came to think about number theoretic problems in connection with probability theory, but I do remember making what appeared to me then to be a great discovery (it was not).

If $\phi(n)$ is the familiar Euler function one has

$$\frac{\phi(n)}{n}=\prod_{p|n}\left(1-\frac{1}{p}\right)$$

[1]See the chapter by Z. W. Birnbaum, pp. 75–82.

which can be written in the form

$$\frac{\phi(n)}{n}=\prod_{p}\left(1-\frac{\rho_p(n)}{p}\right),$$

where

$$\rho_p(n)=\begin{cases}1 p|n\\ 0 p\nmid n\end{cases}.$$

Now, the functions $\rho_p(n)$ are independent and therefore (formally at least)

$$M\left\{\frac{\phi(n)}{n}\right\}=\lim_{N\to\infty}\frac{1}{N}\sum_{n=1}^{N}\frac{\phi(n)}{n}=\prod_{p}M\left\{1-\frac{\rho_p(n)}{p}\right\}$$
$$=\prod_{p}\left(1-\frac{1}{p^2}\right)=\frac{\pi^2}{6}.$$

This, of course, was a well-known elementary fact. But the method also yielded at once

$$M\left\{\left(\frac{\phi(n)}{n}\right)^l\right\}=\prod_{p}M\left\{\left(1-\frac{\rho_p(n)}{p}\right)^l\right\}=\prod_{p}\left\{1-\frac{1}{p}+\frac{1}{p}\left(1-\frac{1}{p}\right)^l\right\}$$

for all l such that the infinite products converge, thus providing a handle on the distribution of $\phi(n)/n$.

When late in November 1938 I left for the United States, the ship (the MS Pilsudksi, subsequently sunk in the early days of World War II) stopped for about six hours in Copenhagen, which gave me a chance to meet Professor Borge Jessen. I communicated my number theoretic discovery to him only to learn that the same result had already been obtained and published by I. J. Schoenberg. The probabilistic nature of the result was, however, somewhat hidden in Schoenberg's proof, and I had the advantage (because of my deep involvement with the normal distribution in unexpected contexts) of being, so to speak, on the ground floor. It was therefore a small step to suspect that the number of prime divisors $\nu(n)$ of n given by the formula

$$\nu(n)=\sum\rho_p(n)$$

should behave like a sum of independent random variables and hence be normally distributed, after subtracting an appropriate mean ($\log\log n$) and scaling down by an appropriate standard deviation ($\sqrt{\log\log n}$). But here my ignorance of number theory proved an impediment. The number of terms in the sum $\Sigma\rho_p(n)$ depends on n, preventing a straightforward application of the central limit theorem. I struggled unsuccessfully with the problem until I stated my difficulties during a lecture at Princeton in March 1939. Fortunately Erdös was in the audience and he perked up at the mention of number theory. He made me repeat my problem, and before the lecture was over he had the proof. Thus did the normal distribution enter number theory and thus was born its probabilistic branch."

Probabilistic number theory is now a thriving branch of pure mathematics, as witness the impressively complete and beautifully written two-volume book by P. D. T. A. Elliott [1].

Statistical independence, once a shadowy partner of gamblers, experimental scientists and statisticians, has achieved the respectability that only an ancient discipline like number theory can bestow. And what about our struggle—mine and my teacher's, Hugo Steinhaus—with independence? If I may be permitted a measure of immodesty, I should like to quote a comment of Hardy on mathematicians of ancient Greece. The Greeks, said Hardy, may appear to us as naïve, unsophisticated, even simpleminded, until a closer look reveals that they were merely fellows of a different college. Well, so were we.

Publications and References

[1] Elliott, P. D. T. A. (1980) *Probabilistic Number Theory* I, II. Springer-Verlag, New York.

[2] Fortet, R. (1940) Sur une suite également répartie. *Studia Math.* **9**, 54–69.

[3] Kac, M. (1938) Sur les fonctions indépendantes (V). *Studia Math.* **7**, 96–100.

[4] Kac, M. and Steinhaus, H. (1938) Sur les fonction indépendantes (IV) (intervalle fini). *Studia Math.* **7**, 1–15.

[5] Kaczmarz, S. and Steinhaus, H. (1936) Theorie der Orthogonalreihen. *Monog. Mat.*, Chapter VII.

[6] Kolmogorov, A. N. (1928) Über die Summen durch den Zufall bestimmter unabhängiger Grössen. *Math. Ann.* **99**, 309–319.

[7] Kolmogorov, A. N. (1929) Bemerkungen zu meiner Arbeit 'Über die Summen Zufälliger Grössen.' *Math. Ann.* **102**, 484–489.

[8] Kolmogorov, A. N. and Khintchine, A. (1925) Über Konvergenz von Reihen deren Glieder durch den Zufall bestimmt werden. *Mat. Sb.* **32**, 668–677.

[9] Łomnicki, Z. and Ulam, S. (1934) Sur la théorie de la mesure dans les espaces combinatoires et son application au calcul des probabilités I. *Fund. Math.* **23**, 237–278.

[10] Rademacher, H. (1922) Einige Sätze über Reihen von allgemeinen Orthogonalfunktionen. *Math. Ann.* **87**, 112–138.

[11] Steinhaus, H. (1922) Les probabilités dénombrables et leur rapport à la théorie de la mesure. *Fund. Math.* **4**, 286–310.

[12] Steinhaus, H. (1930) Sur la probabilité de la convergence de séries. *Studia Math.* **2**, 21–39.

[13] Wintner, A. (1933) Über die statistische Unabhängigkeit der asymptotischen Verteilungsfunktionen inkommensurabler Partialschwingungen. *Math. Z.* **36**, 479–480 and 618–629.

3
MATHEMATICAL STATISTICIANS

Z. W. Birnbaum

Professor Z. William Birnbaum was born in Lwów, Poland on 18 October 1903. After completing his secondary education, he entered the University of Lwów to study first law and then mathematics. He was awarded his LLM in 1925 and his Ph.D. in Mathematics in 1929.

After pursuing mathematical research at Goettingen between 1929 and 1931, he joined an insurance company in Vienna as an actuary for a year, and was later transferred to its offices in Lwów. In 1937 he emigrated to the USA, and obtained an appointment as biometrician at New York University which he held for two years. In 1939, he joined the Mathematics Department at the University of Washington, Seattle; he retired from his position there in 1974 as Professor Emeritus of Mathematics and Statistics.

Professor Birnbaum has taken an active role in the development of statistics in the USA. He has acted as a consultant to Boeing, the Douglas Aircraft Corporation, Rockwell International, and the U.S. Center for Health Statistics among other institutions. He is a Fellow of the American Statistical Association and the Institute of Mathematical Statistics, of which he was elected President in 1963–4. He is a member of the International Statistical Institute, and was editor of the *Annals of Mathematical Statistics* between 1967 and 1970.

Professor Birnbaum was married in 1940 and has two children. He enjoys travel and has been a visiting professor at Stanford, Paris, Rome and the Hebrew University, Jerusalem.

From Pure Mathematics to Applied Statistics

Z. W. Birnbaum

1. Studying in Poland

The city of Lwów in which I was born in 1903, and in which I grew up, was also known as Lemberg when it belonged to the Austro-Hungarian empire. After World War I, in 1918, it became part of Poland. It was there, during my last year in high school (Gymnasium) that I had my first contact with some exciting aspects of mathematics. Our mathematics teacher was a young doctoral candidate with a contagious enthusiasm for his subject. He kept telling us about the wonderful ideas of set theory and topology, the strange things that could happen in the realm of functions of real variables. At that time and place, about 1920 in Lwów, these topics were at the center of attention of a very active group of people who formed the nucleus of what later became known as the Polish School of Mathematics.

After graduating from the gymnasium in 1921, I planned to enroll at the University of Lwów and, guided by the unanimous opinion of my family, study for a degree in something "practical." Mathematics was not considered practical, since it was not of much use in earning a living. Having applied for admission to medical school and the school of engineering and being refused because of the very strict *numerus clausus* (quota system) for Jewish students, I finally enrolled in law school and started attending both law classes and classes in mathematics.

My first mathematics course, "Calculus for Scientists," taught by Professor Ruziewicz, a real variables man, was also my first confrontation with pure mathematics. The opening lectures were devoted to Dedekind cuts as an introduction to real numbers—something I thought I already knew; intuition was played down, abstract rigor was emphasized. I soon got used to this abstract approach (on second thoughts, it served me well in the future) and after only one semester I was invited to take part in mathematics seminars. My parallel studies in law and mathematics continued for four years, the time needed to graduate from law school. I did quite well in law and during my third and fourth years was given an assistantship in constitutional law, possibly as a result of the spectacular success I had in a law seminar by reporting on a paper by G. Pólya which dealt with the mathematics of proportional elections.

In 1925 I graduated from law school with the title *Magister Utriusque Juris*, the "*utriusque*" indicating that I was supposed to be trained in both the secular law of Poland and the canon law of the Catholic church. I joined a law office as a "concipient" (law clerk) and proceeded to spend every free moment in lectures and seminars in mathematics offered by Banach, Steinhaus, Kuratowski, Kaczmarz, Nikliborc and Stozek.

2. School Teaching and Early Research

In 1926 I passed the examinations for a teaching certificate, gave up the practice of law, and found a job teaching mathematics in a gymnasium. My first assignment was a coeducational class of lively 12-year-olds. Convinced, as I was, that mathematics would interest them very much, I followed in my teaching the example of some of my distinguished university teachers: I spoke softly to the blackboard, concentrated on developing ideas, stopped when a proof seemed not to come smoothly. During the very first class period I found myself the target of paper airplanes and spitballs; giggling and talking behind my back drowned out my discourse. I had created what educators call a "discipline problem." Somehow I survived that year, was transferred to another school and there was taken in hand by an experienced older teacher. He taught me a few basic principles such as "always prepare your classes carefully so you don't have to stop to think" and "never turn your back on the enemy." It took me some time, but I recovered from the original trauma, and was later considered a reasonably good teacher.

Meanwhile, the stimulating mathematical atmosphere in Lwów was gradually increasing its hold on me. The daily, and sometimes nightly sessions in the Café Roma, and later on in the Scottish Café (Kawiarnia Szkocka) turned into addictive intellectual feasts. Problems were raised, entered in the problem book which was kept by the head waiter (it was later published as the *Scottish Book*), scribbled on marble-top tables (in pencil—ink was forbidden), and often solved. The author of a solution was entitled to a prize which had been offered and entered in the book, and which ranged from as little as a portion of ice-cream to as much as a bottle of champagne, depending on the estimated difficulty of the problem and the generosity of the sponsor.

While still a law student, I obtained my first new result in analytic functions and showed it to Banach who promptly declared it to be nonsense. A day later, he apologized and asked me to submit it for publication [2]. In 1929, I completed a dissertation on univalent functions of a complex variable, submitted it for publication [3] and in the fall of that year rented the mandatory black suit with long tails and a starched shirt and went through the ceremony of being "promoted" by Professor Steinhaus to Doctor of Philosophy in Mathematics.

I had been saving my teacher's salary to be able to spend some time at the University of Goettingen, in those days the Mecca and Medina of mathematics. The day after receiving my Ph.D. degree I boarded the train for Goettingen, where I was to spend the academic years 1929–30 and 1930–31. There, because of my previous work on analytic functions, I fell into the orbit of Edmund Landau. After a while, I began to drift into other topics such as Sturm–Liouville theorems for differential equations [5], and approximation theorems in function space [4], [25]; in 1931, also jointly with W. Orlicz, I wrote a paper [26] which initiated a line of research continued by others to the present time (e.g. [37]).

This, again, was an exciting mathematical environment. Hilbert was still teaching, in the last year before his retirement; there were also Courant, Bernays, Emmy Noether, Felix Bernstein, Herglotz, Carl Friedrich Siegel, and many others. In addition, spending a few months at a time in Goettingen were visitors from other countries, such as Alexandrov, Kolmogorov, Coxeter, Leo Zippin, to mention just a few.

At the beginning of my second year in Goettingen, Professor Landau offered me an assistantship, but warned me that there was no future in Germany since he was not sure how long even he would be allowed to keep his chair. He was right: two years later the Nazis gained control and Landau had to leave. I took his advice seriously and, to prepare for contingencies, I enrolled in a curriculum leading to the diploma of a life insurance actuary and completed it early in 1931. The director of the university division which offered this professional training was Felix Bernstein, and this gave me an opportunity to become acquainted with him. Work for the actuarial degree was my first serious contact with statistics and its practical applications.

3. Working as an Actuary

In June 1931, I landed a job in the actuarial department of the Phoenix Life Insurance Company in Vienna; this was an international concern with subsidiaries all over the globe. I started working there on the same day as Eugene Lukacs. We were given desks facing each other and were both assigned to E. Helly who was to be our preceptor. This brilliant mathematician (Helly's theorems are classics in functional analysis) was a gentle, warm and cheerful human being, who made my stay at the Phoenix Life Insurance Company enjoyable and instructive in many ways. He showed me over and over again how mathematics could be used to look at a specific actuarial transaction, give it a general formulation, obtain a general solution, and make it part of an inventory of techniques to be used in similar situations in the future. Nowadays, one would carry out the last step by writing a computer program.

In 1932 I was appointed chief actuary of the Polish subsidiary of the Phoenix Life Insurance Company. To my pleasure, this meant moving back to Lwów. There a number of younger people had joined my former teachers: Schauder, Ulam, Sternbach, Kac, Auerbach, Schreier among others, and the outlook for mathematical activity was very bright. At the same time, however, the political climate was turning sinister. The threat of Hitler's Germany was becoming increasingly real and immediate, and Europe's economy was deteriorating. Matters were brought to a crisis for me in 1936 when the Phoenix Insurance Company of Vienna, undermined by political developments, declared bankruptcy. The Polish subsidiary was placed in receivership, the Insurance Commissioner installed a caretaker-manager and I was retained as chief actuary for the transition period until some other company could take over. It became clear that there was little to look forward to, and I decided to emigrate.

4. Emigration to the USA

The United States quota for immigrants from Poland at that time was booked for decades into the future. Fortunately, a leading Polish daily newspaper, the *Ilustrowany Kurjer Codzienny*, appointed me, of all things, their correspondent in the USA. I was given a visitor's visa, and in June 1937 landed in New York.

After a couple of months of writing news reports from the United States, I again met Professor Bernstein, under whom I had worked in Goettingen. He had left Germany and was now Professor of Biometry at New York University. A research assistantship in his department was open; he offered it to me and I gladly accepted. From now on I spent much time on problems such as the relationship of age and presbyopia (old age far-sightedness), and the role of competing risks (a term not known to me at that time) in interpreting the effects of carcinogenic agents [1].

In the autumn of 1937 I started attending Harold Hotelling's statistics seminar at Columbia University. Early in 1939 he suggested that I apply for a faculty position at the University of Washington in Seattle. I applied, was interviewed in New York (at the request of the University of Washington) by the president of the New School for Social Research and by an executive of an oil company, and with the help of a generous letter of recommendation from Albert Einstein was given an appointment as assistant professor of mathematics. The chairman of the department asked me, however, to work on my accent so that "even freshmen could understand" me. I enrolled for summer courses at the University of Vermont, spent the summer practising my English and swimming in Lake Champlain, and since then have had no difficulty in being understood even by freshmen.

5. The University of Washington

At the University of Washington, besides teaching courses in mathematics, I started developing a curriculum in mathematical statistics and adding statisticians to our group within the Mathematics Department. In 1948 I received a basic research contract with the U.S. Office of Naval Research and this helped our group to achieve more formal recognition, under the title of "Laboratory for Statistical Research." This laboratory kept increasing and at times consisted of 12 full-time faculty members as well as assistants, research assistants and clerical staff. Their work ranged from probability theory and theoretical statistics to applications, although our position in the Mathematics Department made it necessary to emphasize the mathematical aspects of our activities.

It was only as recently as 1978 that the administration of the University of Washington authorized a separate Department of Statistics. This now includes statisticians who were formerly in the Mathematics Department and several statisticians holding joint appointments with other departments.

The research contract with the Office of Naval Research was continued for 25 years, from 1948 until my mandatory retirement from teaching in 1974. Research under this contract as well as under grants from the Guggenheim Foundation and the National Science Foundation was often conducted jointly with other members of our statistics group. Most of them were graduate students who went on to develop their own fields of interest and earned individual recognition as research scholars. I enjoyed these years of teaching and working with younger people. Looking back on my former students and fellow workers I have learned to appreciate the wisdom of my teacher Hugo Steinhaus, who used to say that there is no greater joy for a teacher than to have taught people who became better than he.

Our work dealt with probabilistic inequalities [6], [23], [29], [36], with effects of pre-selection or truncation [8], [9], [16], [17], [24], [27] and of non-response in sampling surveys [32], with properties of Kolmogorov–Smirnov-type and related statistics [11], [21], [22], [28] and numerical tabulations of these statistics for small sample sizes [10], [35], and with the use of computers in unconventional situations [13], [14].

A research project, in cooperation with staff members of the Boeing Scientific Research Laboratories, begun about the year 1958 and continued for a decade, was devoted to the statistical theory of life-lengths of materials [19], [30], [31] and of the reliability of multicomponent systems [12], [18], [20].

Work with the U.S. National Center for Health Statistics led to new multiple-sampling designs [33] and, recently, to the publication of a monograph on competing risks [15]. A study of biases and variances of infant mortality rates has been completed and should be published soon [34].

6. Concluding Remarks

In the course of years I have served as consultant for airplane manufacturers and for companies engaged in managing radioactive waste, for government agencies, as expert witness in law suits, as editor of professional journals, as consulting editor to publishers, and have found this variety of activities stimulating and exciting, with a strong appeal to my sense of adventure.

In some instances an applied program led to an abstract formulation, then to a theoretical solution, then in turn to a rephrasing of the results in terms of their practical use, and sometimes to a numerical tabulation which could be placed in the hands of the practitioner who had originally stated the problem. In many such ways, theory and applications often challenged, supported and advanced each other. In the long run, I had the good fortune, in my teaching and my other work, to move in the direction suggested in the title of this essay: from the rarefied atmosphere of pure mathematics to the invigorating environment of applicable statistics.

Publications and References

[1] BERNSTEIN, F., BIRNBAUM, Z. W. and ZACHS, S. (1939) Is or is not cancer dependent on age? *Amer. J. Cancer* **37**, 298–311.

[2] BIRNBAUM, Z. W. (1927) Quelques remarques sur l'intégrale de Cauchy. *Arch. Soc. Sav. Sci. Lett. Leopol.* **4** (10) 3, 1–15.

[3] BIRNBAUM, Z. W. (1929) Zur Theorie der schlichten Funktionen. *Studia Math.* **1**, 159–190.

[4] BIRNBAUM, Z. W. (1930) Über Approximation im Mittel II. *Goettinger Nachr. Math.-Phys. Klasse*, 338–343.

[5] BIRNBAUM, Z. W. (1930) Abschätzung der Eigenwerte eines Sturm–Liouvilleschen Eigenwert-Problems mit Koeffizienten von beschraenkter Schwankung. *Arch. Soc. Sav. Sci. Lett. Leopol.* **5** (2) 7, 1–10.

[6] BIRNBAUM, Z. W. (1942) An inequality for Mills' ratio. *Ann. Math. Statist.* **13**, 245–246.

[7] BIRNBAUM, Z. W. (1948) On random variables with comparable peakedness. *Ann. Math. Statist.* **19**, 76–81.

[8] BIRNBAUM, Z. W. (1950) Effect of linear truncation on a multinormal population. *Ann. Math. Statist.* **21**, 272–279.

[9] BIRNBAUM, Z. W. (1950) On the effect of the cutting score when selection is performed against a dichotomized criterion. *Psychometrika* **15**, 385–389.

[10] BIRNBAUM, Z. W. (1952) Numerical tabulation of the distribution of Kolmogorov's statistic for finite sample size. *J. Amer. Statist. Assoc.* **4**, 425–441.

[11] BIRNBAUM, Z. W. (1953) On the power of a one-sided test of fit for continuous probability functions. *Ann. Math. Statist.* **24**, 484–489.

[12] BIRNBAUM, Z. W. (1968) On the importance of components in a system. *Selected Statistical Papers* 2, Mathematical Centre Tracts **27**, Amsterdam, 83–95.

[13] BIRNBAUM, Z. W. (1974) Computers and unconventional test-statistics. In *Reliability and Biometry*, SIAM, Philadelphia, 441–458.

[14] BIRNBAUM, Z. W. (1975) Testing for intervals of increased mortality. In *Reliability and Fault Tree Analysis*, SIAM, Philadelphia, 413–426.

[15] BIRNBAUM, Z. W. (1979) On the mathematics of competing risks. U.S.-OHEW Publication PHS 79-1351, I-VI, 1-58.

[16] BIRNBAUM, Z. W. and ANDREWS, F.C. (1949) On sums of symmetrically truncated random variables. *Ann. Math. Statist.* **20**, 458–461.

[17] BIRNBAUM, Z. W. and CHAPMAN, D. G. (1950) On optimum selections from multinormal populations. *Ann. Math. Statist.* **21**, 433–447.

[18] BIRNBAUM, Z. W. and ESARY, J. D. (1965) Modules of coherent binary systems. *J. SIAM* **13**, 444–462.

[19] BIRNBAUM, Z. W., ESARY, J. D. and MARSHALL, A. W. (1966) A stochastic characterization of wear-out for components and systems. *Ann. Math. Statist.* **37**, 816–825.

[20] BIRNBAUM, Z. W., ESARY, J. D. and SAUNDERS, S. C. (1961) Multi-component systems and structures and their reliability. *Technometrics* **3**, 55–77.

[21] BIRNBAUM, Z. W. and HALL, R. A. (1960) Small sample distributions for multi-sample statistics of the Smirnov type. *Ann. Math. Statist.* **31**, 710–720.

[22] BIRNBAUM, Z. W., and MCCARTY, R. C. (1958) A distribution-free upper confidence bound for $\Pr(Y < X)$ based on independent samples of X and Y. *Ann. Math. Statist.* **29**, 558–562.

[23] BIRNBAUM, Z. W. and MARSHALL, A. W. (1961) Some multivariate Chebyshev inequalities with extensions to continuous parameter processes. *Ann. Math. Statist.* **32**, 687–703.

[24] BIRNBAUM, Z. W. and MEYER, P. L. (1953) On the effect of truncation in some or all coordinates of a multinormal population. *J. Indian Soc. Agricultural Statist.* **5**, 17–28.

[25] BIRNBAUM, Z. W. and ORLICZ, W. (1930) Über Approximation im Mittel I. *Studia Math.* **2**, 197–206.

[26] BIRNBAUM, Z. W. and ORLICZ, W. (1931) Über die Verallgemeinerung des begriffes der zueinander konjugierten Funktionen. *Studia Math.* **3**, 1–67.

[27] BIRNBAUM, Z. W., PAULSON, E. and ANDREWS, F. C. (1950) On the effect of selection performed on some coordinates of a multi-dimensional population. *Psychometrika* **15**, 191–204.

[28] BIRNBAUM, Z. W. and PYKE, R. (1958) On some distributions related to the statistic D_N^+. *Ann. Math. Statist.* **29**, 179–187.

[29] BIRNBAUM, Z. W., RAYMOND, J. and ZUCKERMAN, H. S. (1947) A generalization of Tshebyshev's inequality to two dimensions. *Ann. Math. Statist.* **18**, 70–79.

[30] BIRNBAUM, Z. W. and SAUNDERS, S. C. (1958) A statistical model for life-length of materials. *J. Amer. Statist. Assoc.* **53**, 151–160.

[31] BIRNBAUM, Z. W. and SAUNDERS, S. C. (1969) A new family of life distributions. *J. Appl. Prob.* **6**, 319–327.

[32] BIRNBAUM, Z. W. and SIRKEN, M. G. (1950) Bias due to non-availability in sampling surveys. *J. Amer. Statist. Assoc.* **45**, 98–111.

[33] BIRNBAUM, Z. W. and SIRKEN, M. G. (1965) Design of sample surveys to estimate the prevalence of rare diseases. U.S. Public Health Service Publication No. 1000 Serial 2 No. 11, 1–8.

[34] BIRNBAUM, Z. W. and SIRKEN, M. G. (1981) Infant mortality rates as estimates: their biases and variances. Technical Report No. 4. Department of Statistics, University of Washington.

[35] BIRNBAUM, Z. W. and TINGEY, F. H. (1951) One-sided confidence contours for probability distribution functions. *Ann. Math. Statist.* **22**, 592–596.

[36] BIRNBAUM, Z. W. and ZUKERMAN, H. S. (1944) An inequality due to H. Hornich. *Ann. Math. Statist.* **15**, 328–329.

[37] BUND, I. M. (1978) *Birnbaum–Orlicz Spaces.* Instituto de Matematica e Estatistica da Universidade de Sao Paulo.

R. C. Bose

Raj Chandra Bose was born in Hoshangabad, Madhya Pradesh, India on 19 June 1901. He went to school in the town of Rohtak, Haryana; his father, a doctor, expected him to achieve the highest scholastic standards, and Raj rarely disappointed him.

After graduating from school, he entered the Hindu College, Delhi in 1917 and did very well in his studies. His mother died in 1918 and his father in 1920 leaving him as the oldest son of the family of five; he became responsible for his younger brother and two sisters, his oldest sister having married. There followed a period of great hardship which the family was fortunate to overcome, and he was able to graduate with a BA Honors in mathematics in 1922 and an MA in applied mathematics in 1924.

He went to Calcutta in 1925 and obtained a second MA in pure mathematics from Calcutta University in 1927, standing first in his class. He became lecturer in Mathematics at Asutosh College, Calcutta in 1930 and married Sandhya Lata Datta in 1932; they have had 2 daughters, Purabi (born in 1934) and Sipra (born in 1938). It was in 1932 that he met Professor Mahalanobis, who persuaded him to join the Indian Statistical Institute in 1933. As a result, he soon started his research in multivariate analysis and experimental design. He was appointed lecturer in the Department of Statistics at Calcutta University in 1941, and head of the Department in 1945, following Mahalanobis.

In 1949, he left India to accept a Professorship in Statistics at the University of North Carolina, Chapel Hill and worked there until his retirement at 70 in 1971. His 22 years at Chapel Hill proved to be the most productive and happy of his life; during the five years 1966–71, he was the holder of the endowed Kenan Chair. In 1971, he became Professor of Mathematics and Statistics at Colorado State University, Fort Collins, and began another career in his retirement.

Professor Bose has been influential in statistics not only in India and the USA, but also internationally. He has been the recipient of many honors, among them a D. Litt. from Calcutta University in 1947, an honorary D.Sc. from the Indian Statistical Institute in 1974, the Presidency of the Institute of Mathematical Statistics in 1971–72, membership of the U.S. National Academy of Sciences in 1976 and an honorary D. Litt. from Visvabharati, Tagore's university, in December 1979. His hobbies are travel, hiking, gardening, history, art and culture.

Autobiography of a Mathematical Statistician

R. C. Bose

1. Early Life and Schooling

Towards the end of the nineteenth and the beginning of the twentieth centuries many people from Bengal (West Bengal and Bangladesh) emigrated to Western and Central India. Among these émigré families were the families of my mother and the husbands of both my aunts (my father's sisters). My father, the late Protap Chandra Bose, came from the well-known Bose family of Kholsini (near Chandernagar, West Bengal). After serving his term as a military doctor in the British army, he had opened a private practice in the town of Rohtak (Haryana) where his sister's husband was employed at that time. He married my mother Ushangini Mitra during one of his visits to his other sister who was living in Hoshangabad (Madhya Pradesh). This was his second marriage; his first wife died without having children. I was the first child of this union and was born at my mother's home in Hoshangabad on 19 June 1901. I was followed by one brother (1905) and three sisters (1907, 1910, 1914).

I spent my early life in Rohtak and was educated at the Government High School there. My father was a strict disciplinarian and expected me to be at the top of my class in each individual subject. This I usually succeeded in doing. I remember that in the mid-term examination in grade eight, I came out second in geography. My father was very angry and ordered me to learn Sohan Lal's geography of the world (a 200-page book) by heart. I did this to his satisfaction, being able to repeat the whole book almost verbatim. My father had a photographic memory and I have inherited it from him. At that time I did not consider it as exceptional. I fell sick of influenza at the time of the 1917 matriculation examination of the Punjab University, (which comprised the area covered by the present West Punjab, East Punjab, Haryana and Delhi) and narrowly missed a university scholarship. In fact there were 28 scholarships and I was 29th in order of merit.

2. University and a Start in Research

I joined the Hindu College Delhi in April 1917. My early years in college were one of the best times in my life. New horizons in mathematics and science which I loved were opening up. The study of Sanskrit literature was also a pleasant experience and I can still repeat from memory substantial portions of the Kumar Sambhava of Kalidas. I also got an introduction to the great heritage of ancient Hindu religion and culture in the study of the Geeta and the Upanishads, which were taught on a voluntary basis. This idyllic state of affairs came to a sudden end with the death of my mother in October 1918, during the great influenza epidemic which swept the world towards the end of the First World War.

The 1919 Intermediate examination of the Punjab University was postponed due to political disturbances: this was the year of the Jalianwala Bagh massacre. When the examination was finally held, I managed to stand first in the university, which was a source of great satisfaction to my father, though he was never the same man after my mother's death. He did not long survive her, and died of a stroke in January 1920. This left us orphans. My brother was still in school in Rohtak. I left him and my two younger sisters with my cousins who agreed to provide them board in exchange for rent-free occupation of our house at Rohtak which my father had built. A marriage had already been arranged for the eldest of my sisters by my father, just before his death. In the summer of 1921, after my brother had completed school, I moved my family to Delhi, where the four of us lived in a single room provided by the family of a student whom I was coaching. My brother started college and took a coaching job. With this income and the scholarships which both of us were getting, we somehow managed to survive. My BA Honors examination (in 1922) was two months away when our coaching jobs expired. My scholarship was also due to expire. We were on the brink of starvation on the eve of the examination. Fortunately, we could sell a bottle of quinine from the stock of medicine left by my father. This fetched a good price as quinine was very scarce in the aftermath of the First World War. Still, prospects looked very bleak, for even if I won a scholarship, I would not start to receive it until the results were declared. We were, however, providentially saved. I got a message from the headmaster of the St. Stephens High School, where a temporary one-year teaching position in mathematics was open, to attend an interview with him. He asked me to join the day after the examination, as I had an excellent recommendation from our principal. I taught there for one year. Also, I duly won the scholarship and was allowed to join the MA course in Applied Mathematics even though I could not attend all the classes because of my school job. In 1923 my brother passed the Intermediate examination and won a scholarship. After my school job was over in April 1923, I continued my MA studies and we managed with our scholarships and by coaching students. As

I had not been able to devote full time to my studies, in the MA examination (1924) (Applied Math.) of Delhi University (which had separated from Punjab University in 1921) I could not get a first class, even though I had obtained the first place. I had hoped for a position in Delhi in some college but I was unable to secure one. For one year we had to subsist on our income from coaching. My brother passed the BA in 1925 and got a teaching job in St. Stephens High School where I had previously taught.

Now my fortunes took another turn. Seth Kedarnath Goenka, whose younger brother I was coaching, was so pleased with my work that he agreed to support me for as long as I needed, in Calcutta, where I wanted to study pure mathematics. I thus came to Calcutta in 1925. My sisters remained with my brother, who now had a steady job. Here I had the good luck to catch the attention of Professor Shyamadas Mukherjee, a very fine geometer of the old school. His main interests at that time were non-Euclidean geometry, n-dimensional geometry and global (*im grössen*) properties of convex curves. He gave me a room in his house and free use of his library. He also secured a good coaching job for me so that I no longer needed any help from Sethji. I stood first in the first class in the MA examination in Pure Mathematics (1927), winning the University and the Keshalilal Mallik gold medals. I was research associate in mathematics from January 1928 to December 1929, working in geometry under the guidance of Dr Mukherjee. I continued to live at his house. He was a dedicated teacher; from him I learned the spirit of mathematical research. I also made good use of his library and learned all the mathematics I could. My special favorites were Hilbert's *Grundlagen der Geometrie* and Klein's *Icosahedron* and *Elementary Mathematics from a Higher Point of View*. The stimulus thus provided has had a profound influence on my mathematical career. I also arranged a marriage for my second sister in 1928.

3. A Lectureship in Mathematics: The Indian Statistical Institute

In 1930 I was again faced with the problem of finding a job. In the wake of the great stock-market crash in the United States, there was a world-wide depression and nothing was to be had. In most places I was turned away with the remark that I was overqualified for the position they had. Ultimately, in March 1930 I found a position as a lecturer in mathematics in the Asutosh College, Calcutta. As the salary was very low I had to supplement it by coaching students for the MA or BA Honors in mathematics or for competitive examinations for government service. Until then, I had been living with Dr Mukherjee. Now I rented a small apartment and brought my youngest sister from Delhi, freeing my brother. My second sister, who now had a daughter, came to live with me for a few months to set up the house. I kept up some research, especially in non-Euclidean geometry, and differential geometry of convex curves [1] under heavy odds, as my teaching and coaching duties left me with little time.

In September 1932 I married Sandhya Lata Datta, daughter of Kamini Kumar Datta, a retired district judge. Since then she has been my constant companion and helper, and whatever I have achieved is in large part due to her.

One morning in December 1932, my servant told me that a "sahab" wanted to see me. He had come in a car. In those days, owners of cars were very important people indeed, and "sahab" was a very honorable designation. It turned out that the person who wanted to see me was the famous Professor P. C. Mahalanobis, head of the Department of Physics in the Presidency College, Calcutta. He had founded the Indian Statistical Institute in 1931. During his studies in Cambridge and London, he had come under the influence of Karl Peason and R. A. Fisher, the great pioneers of statistics in England, and was now looking for suitable people to advance the study and application of statistics in India. My geometrical work had come to his attention. I told him that I had only a rudimentary knowledge of statistics, as statistics was not my speciality. "You will learn it," he said, and meanwhile he was willing to give me a half-time job in the Institute. I was to work for the Institute on Saturdays, and for the summer and Pujah vacations. (The university and colleges in Calcutta used to have two main vacations during the year: a seven-week summer vacation beginning approximately during the third week of April, and a five-week Pujah vacation beginning approximately during the first week of October. Mahalanobis usually moved to some hill town resort during the summer vacation with a selected staff and to Giridih (in the Santhal Parganas in Bihar) during the Pujah vacations.) I thus joined the Indian Statistical Institute in January 1933, as a research scholar.

During the summer of 1933 we moved to the resort town of Darjeeling, where family accommodation was provided for me. The day after I had settled down, the secretary brought me the volumes of *Biometrika* (from 1900–1932), the most important statistical journal at that time, together with a typed list of 50 papers. There was also R. A. Fisher's book *Statistical Methods for Research Workers*. When I went to see Mahalanobis, he said, "You were saying that you do not know much statistics. You master the 50 papers, the list of which you have received, and Fisher's book. This will suffice for your statistical education for the present." Thus began my education in statistics. A few months later Mahalanobis also recruited S. N. Roy, an Applied Mathematics MA, on the same terms as myself. He was to become one of my best friends and collaborators.

S. N. Roy and I were the two chief mathematicians in the Institute. Mahalanobis would select one statistical topic every summer. It was Roy's and my duty to master it, read all the important papers, and give lectures to the other junior members of the staff. Thus, both of us became well grounded in statistics in a few years.

4. Mathematics and Statistical Research

Statistical theory was in an interesting situation during the 1930s. Methods had been devised for the case when only a single character, like the yield of some crop, was under study. But methods were being developed for "multivariate analysis," the case when each individual was to be studied with respect to a number of characters. Mahalanobis was interested in the study of differences between various races, and for this purpose had devised a statistical measure called the Mahalanobis distance or D^2, in the distribution of which he was extremely interested. He had himself, by very laborious methods, discovered the first four moments for the classical case, i.e., when the population variances and covariances are exactly known [26]. In 1934, while at Giridih, I succeeded in discovering the correct distribution for the classical case, which turned out to involve an I-Bessel function of purely imaginary argument [2]. Mahalanobis was now constantly pressing us to obtain the distribution of D^2 in the "studentized" case, i.e., when the population variances and covariances had to be estimated from the samples themselves. This, however, seemed to be out of reach until the Wishart distribution of estimated variances and covariances could be translated into a manageable form. Using Fisher's idea of representing an n-sample by a point in a Euclidean space of n dimensions, Mahalanobis, Roy and myself introduced the idea of t-coordinates or "rectangular coordinates." Though in our original paper we defined them geometrically, they can be easily expressed in a matrix form. If $S=(s_{ij})$ is the covariance matrix of the estimated variances and covariances, $T=(t_{ij})$ is an upper-triangular matrix, i.e. $t_{ij}=0$ when $i>j$, then $T'T=S$. Again if Σ is the population covariance matrix and Δ is an upper-triangular matrix such that $\Delta\Delta'=\Sigma^{-1}$, then $L=T\Delta$ is again an upper-triangular matrix. We called l_{ij} the normal coordinates. For a p-variate population the $p(p+1)/2$ normal coordinates turn out to be independently distributed. For $i\neq j$ $n^{1/2}l_{ij}$ is normally distributed, and $n^2 l_{ij}$, $i\neq j$ has the χ^2 distribution with $n-i+1$ degrees of freedom [8]. Using these results and Fisher's geometrical method, Roy and I were then able to obtain the distribution of the "studentized D^2" [12], which in essence is the non-central distribution of Hotelling's T^2 [24].

Meanwhile S. N. Roy and myself had joined the Institute in a full-time capacity in 1935. In 1936 there was a development that was to influence my career profoundly. Dr Ganesh Prasad, the head of the Department of Mathematics in the University of Calcutta, had passed away, and Professor F. W. Levi, who had left Germany during the Nazi regime, became the Hardinge Professor and chairman of the department. I used to attend his seminars on algebra and geometry and became especially interested in "finite fields" and "finite geometries" which were quite new to me. The professors in the mathematics department shirked the effort to learn the new algebra. Thus, Levi offered me a half-time position in the University to teach "modern algebra" in 1938.

R. A. Fisher, the leading statistician in the world at that time, came in 1938–39 as a visiting professor to the Institute. At that time he was interested in the study and construction of patterns called statistical designs, to be used in statistically controlled experiments. He posed to us some problems arising out of this study. It occurred to me that I could successfully use the theory of finite fields and finite geometries, which I was then teaching, as tools for the construction of designs. I solved during his stay one of the problems he had posed. He asked me to develop my methods systematically and send him a paper which he promised to publish in the *Annals of Eugenics* of which he was editor. Thus my paper on "The Construction of Balanced Incomplete Designs" came to be written in 1939; it has now become a classic and is quoted in every book on the subject [3].

In collaboration with Nair, I discovered a very general type of designs which we named partially balanced designs [10], further work on which will be described later.

The fortuitous combination of circumstances described above changed the direction of my work and I was to devote the next 18 years of my life (1940–58) almost wholly to the study of various aspects of statistical designs. During this period, my students and I were among the main contributors to this area of statistical theory.

In 1940 I briefly left the Institute and joined the University of Calcutta as lecturer in the Department of Mathematics (January 1940–June 1941). In 1941 the University of Calcutta decided to start a postgraduate Department of Statistics, awarding MA and M.Sc. degrees in statistics. S. N. Roy and I became the first lecturers in the new department, with Mahalanobis as the honorary head. I also rejoined the Institute in a part-time capacity. Most of the work of organizing the new department and setting up courses of study fell on my shoulders. Among the first batch of students was C. R. Rao, who later became the director of the Indian Statistical Institute after Mahalanobis's retirement, and is now a world-famous statistician.

Besides my work on incomplete block designs, the main work which I did was to develop jointly with K. Kishen [7] a general theory of confounding for orthogonal symmetrical factorial experiments, in which each treatment was represented by a point in affine finite space, as an alternative to Fisher's work using group theory for the same purpose [22]. In 1947 I succeeded in generalizing Fisher's theorem answering the question: what is the maximum number of factors which, given the number of levels and replications, can be accommodated in a symmetrical factorial experiment so that no t factor interaction is confounded [4]. The case $t=2$ is due to Fisher [23].

In 1945, Mahalanobis relinquished the headship and I became the head of the Department of Statistics in the University of Calcutta, and continued my part-time work in the Institute. Though my research was well recognized at this time, I had not taken the doctor's degree. The university authorities told me that to get a full professorship this situation had to be rectified. I therefore submitted my published papers, together with an introduction reviewing my work in multivariate analysis and design of experiments, as a doctoral thesis for the degree of D. Litt. in 1947. Among my "examiners" was R. A. Fisher. I duly got the degree.

5. A Broadening of Interests

I have detailed my academic work and career from 1930–1947. Before continuing the story, I shall pause here and describe some of my other activities and interests during this period. In the struggle for existence after the death of my father and before the completion of my formal education in 1927, I had no time to pursue other interests. However, in 1928, when I was still a research scholar and living with Dr Mukherjee, I developed an interest in Bengali literature, especially the works of Rabindranath Tagore. During the next three years I read all his books and became a fairly good scholar of Tagore. This interest was further heightened when I joined the Institute, for Mahalanobis, himself a man of many accomplishments, was well versed in Rabindric literature, and was a personal friend of the poet. Tagore, when he came to Calcutta, often stayed at Mahalanobis's house, and we, the members of the Institute, had free access to him.

I have already described how we generally went to some mountain stations during the summer and to Giridih in autumn. I developed an interest in hiking. During my years at the Institute, my wife and I, either alone or accompanied by others, took many long trips in the Himalayas around Darjeeling, Kalimpong, Mussorie and Simla.

The Science Congress had its annual meeting in Madras in early January 1940, so at Christmas 1939, my wife and I, accompanied by my colleague, K. Kishen, decided to take a tour of southern India, and in particular visited Mahaballipuram, Ramesweram, Dhanushkodi, Madura, Trichnapaly, Chidambaram and Tanjore. I enjoyed this tour so much that it became a regular ritual over Christmas to see the country around any place where the Science Congress was to be held in January. In this way, my wife and I have seen all the various parts of India. I have already indicated my interest in Indian religion and philosophy but my travels now gave me an interest in Indian art, architecture and history.

In January 1947, I received an invitation from Columbia University, New York, and the University of North Carolina at Chapel Hill to teach there as a visiting professor during the fall and spring semester of 1947–48. Shortly afterwards the Virginia Polytechnic Institute at Blacksburg invited me to give a course during the second summer term of 1947. Thus I went to the United States in early August, on the eve of Indian independence. Some of the Indian students at Blacksburg arranged for a celebration on 15 August, Indian Independence Day, which was attended by many Americans, including the president of the university. I was the principal speaker. Before joining Columbia in late September, I also attended a meeting at Woods Hole as the representative of India, where the International Biometric Society was founded. Here I met many important American statisticians. During my stay at Columbia and Chapel Hill I visited many famous American universities and lectured on my own work and also the work of my other colleagues at the Indian Statistical Institute. I liked the atmosphere in the American universities very much, including the comparative freedom from internal politics. Before leaving Chapel Hill in May 1948, I received an offer from the university to accept a full professorship in a permanent capacity. I also had a similar offer from the University of Illinois at Urbana. I also received an offer from the University of California at Berkeley to teach there during the summer of 1948. On my way from Chapel Hill to Berkeley I chose a roundabout route and visited the most important tourist sights like Niagara Falls and the Grand Canyon. I also lectured at Chicago and Iowa. Mahalanobis was visiting Chicago at that time and offered me the assistant directorship of the Institute of Statistics after my return to India. I returned to India at the end of August 1948. Professor Levi had retired, and on his recommendation I was also offered the Hardinge Professorship and the position of Head of the Department of Mathematics which went with it, in the University of Calcutta. I thus had many offers to choose from.

6. A Research Career in America

People have always asked me why I decided to go to America. I was already 47 years old. The two positions offered to me in India involved a lot of administrative work which would have meant the end of my research career. I opted for research and joined the University of North Carolina at Chapel Hill, as Professor of Statistics, in March 1949. I have never regretted this decision. The 22 years that I spent at Chapel Hill have been some of the most fruitful and happy years of my life. S. N. Roy joined me at Chapel Hill a year later, in April 1950. We were not completely cut off from India. Both of us encouraged bright Indian students to apply for fellowships which, of course, were awarded by our department strictly on the basis of merit, without distinction of race or nationality. While I was at Chapel Hill seven Indian students completed their Ph.D. dissertations under my direction, including S. S. Shrikhande who has just retired from the headship of the Department of Mathematics at the University of Bombay. Many of my students now occupy distinguished academic positions in India, the United States, and other parts of the world as well. I have already described some of my work in the design of experiments; I continued this in Chapel Hill.

The definitions and analysis of partially balanced designs, as given in the Bose–Nair paper of 1939 [10], were rather difficult and had not been well grasped by the statistical public. After coming to Chapel Hill I realized that special cases of them were being discovered by various authors. I therefore decided to simplify the definition and to classify them with the help of my students. The first step in this direction was to introduce the idea of an association scheme [13]. If the treatments formed an association scheme then we could base a partially balanced incomplete block (PBIB) design on it if certain further conditions were satisfied. The simplest type of association scheme is the group divisible. The combinatorial properties of group divisible designs were studied in [6], and the problem of constructing them in [16]. Mesner and I [9] studied the linear associative algebras corresponding to association schemes of PBIB designs. The concept of association schemes has proved useful in other branches of mathematics, for example group theory and coding theory. It is interesting to observe that most of the designs obtained by my students and myself were inspired by Fisherian principles, but important subclasses of them have turned out to be optimal in the sense of optimal design theory. Though optimal design theory goes back much further, Kiefer [25] and his associates have systematically studied it in recent times. Cheng in particular has demonstrated the optimality of many group divisible designs and PBIB designs [19], [20].

In 1955, I developed an interest in coding theory, which culminated in the discovery (in collaboration with my student Ray-Chaudhuri) of the now well-known Bose–Chaudhuri codes (discovered independently by Hocquengham) [11]. A code is a means of transmitting information over a channel. At the end of each message sequence of symbols (or interlaced with it) one adds a sequence of redundant symbols calculated from the message in such a way that if there occur less than a prescribed number of errors during transmission, they can be corrected at the receiving end. It turned out that an error-correcting code was very deeply connected with the theory of confounding which I have already described, especially with my generalization of Fisher's theorem.

The same year (1959) I also made another famous discovery in collaboration with my student Shrikhande who was a visiting professor in our department that year. The great Swiss mathematician, Leonhard Euler, had in 1782 made a conjecture the truth or falsity of which was undecided for more than 175 years. Shrikhande and I proved his conjecture wrong.

Consider a square with v^2 cells, there being v cells in each row and each column. If there is a set of n symbols, filling the cells so that each symbol occurs exactly once in each row and each column, then we have a Latin square. Two Latin squares are said to be orthogonal if when superposed any particular symbol of the first square occurs exactly once with each of the v symbols of the second square. Euler showed that we can always construct two orthogonal Latin squares of any order v, if v is not of the form $4t+2$. He tried very hard to find two orthogonal Latin squares when the length of the side was 6 but failed. He made the conjecture that two orthogonal Latin squares cannot be constructed when the length of the side is a number which when divided by 4 leaves a remainder 2; for example, 6, 10, 14, 18, 22,... [21]. Tarry proved in 1900 that at least for 6 Euler was right [29]. But the general question was undecided until Shrikhande and I showed that Euler's conjecture was false for 22 by actually constructing two orthogonal Latin squares with side 22 [14]. We went on to construct squares of this type for many other numbers [15]. Parker also found new methods for constructing other squares of this type [27]. Finally, in 1960, the three of us proved that Euler was wrong in every case when the length of the side exceeds 6, and leaves a remainder 2 when divided by 4 [17].

This reminds me of an interesting incident. Our paper was read at the New York meeting of the American Mathematical Society. The science editor of the *New York Times* came to interview us, and next morning our picture appeared on the first page of the Sunday edition of the *New York Times* with a description of our work. I was staying at a small hotel. Next morning when I went to pay the bill, the cashier looked at me and asked, "Is it your picture in the *New York Times*?" "Yes, it is my picture," I said. He replied, "You must have done something. The front page of the *New York Times* cannot be bought for a million dollars."

During the years 1958 to 1963 I devoted myself mainly to coding theory and asymmetric factorial designs [18].

The type of work that I had hitherto done belongs to a branch of mathematics called combinatorics, which deals with discrete and finite mathematical structures. Owing to the rapid development of computer technology and information theory after 1950, combinatorics received a great impetus. Up to 1964, my work had been mainly motivated by applications to statistics and coding theory. Though I continued some statistical work, I now became interested in other combinatorial problems in general for their mathematical interest, apart from any applications which they might have. In particular, I became interested in the interconnections between the structure of "designs" and "graphs." This work I am still continuing, but it is too far from probability and statistics to be of much interest for the readers of this article [5]. In a way things have come full circle. I started out as a mathematician and, after devoting 30 years to statistics, have come back to mathematics. My students, Shrikhande and Ray-Chaudhuri, have also made notable contributions to this field and have received international recognition. Another student of mine, J. N. Srivastava, has carried forward my work on designs and is now one of the foremost authorities on the subject.

7. Retirement and a New Career

In August 1971, I retired from the University of North Carolina at Chapel Hill, at the age of 70. During my last five years (1966–71) at Chapel Hill I was Kenan Professor, which is an endowed chair. On my retirement from North Carolina I was offered the position of Professor of Mathematics and Statistics at Colorado State University in Fort Collins, from which I retired in September 1980. I am continuing as Emeritus Professor and still have a Ph.D. student working under me. When he completes his work, he will be my thirtieth Ph.D.

Many honors and awards have come my way during recent years. In September 1971, an International Symposium on Combinatorics was held in Fort Collins in honor of my seventieth birthday and the proceedings presented to me as a Festschrift [28]. The Association of Indians in America selected me as one of the five people to receive recognition at their first Honor Banquet in 1973. The Indian Statistical Institute gave me an honorary D.Sc. degree in December 1974. I was especially gratified by the presence at each of these functions of many of my old students. In 1971–72 I served as the President of the Institute of Mathematical Statistics. In 1976 I was elected a member of the U.S. National Academy of Sciences. This is the highest honor an American scientist can receive. In 1979 I was one of the 14 Indian scientists and men of letters to receive an award from the Indian Ambassador at an honor banquet held at the Hilton Hotel. In the same year I was awarded an honorary degree of Desikottama (D. Litt.) by the Visvabharati, the international university founded by Tagore.

I shall now describe some of the other details of my life in the United States. During my stay in Chapel Hill, I developed an interest in gardening. Most of the flowers, shrubs and trees in the yard of our house at Chapel Hill (built in 1952) were planted and cared for by my wife and me. We had a beautiful garden admired by everybody.

During the years 1949–1963, my wife and I traveled all over the United States and Canada and visited almost all the National Parks and other places worth seeing. Of course, this travel was coordinated (as earlier in India) with scientific conferences which I attended and where I presented papers. In 1955 I was visiting professor at University College London and the London School of Economics. During the summer of that year, accompanied by my wife and two daughters, I traveled by car through France, Germany, Austria, Switzerland and Italy. This was the first of our many travels in Europe. I must mention here that I have never learned to drive. All the driving is done by my wife. In 1961 my wife and I visited India and stopped in Egypt for three weeks lecturing and touring. In December I went as visiting professor to the University of Geneva. We traveled in Germany and Denmark, paying a visit to Professor and Mrs Levi who had settled in Freiburg after retirement. From 1965 onwards my wife and I have visited Europe almost every summer. In 1978 I was visiting professor at the

Institute of Geometry in Bologna, Italy. Before arriving at Bologna I lectured at the University of Stuttgart and the University of Geneva, and traveled in Germany and Switzerland. After Bologna, my wife and I spent two weeks traveling in Norway, then attended the International Mathematics conference in Helsinki. We then proceeded to Hungary where I was an invited speaker at the International Symposium on Algebraic Graph Theory held at Szeged. This was my sixteenth visit to Europe. My travels have given me an abiding interest in the history, art and culture of Europe and America. I have visited almost all of the famous art galleries of the world. I am now a fairly good amateur historian and art critic and have a very good collection of art and history books, most of which I have read.

8. Concluding Remarks

I shall conclude with a few details of my family life. I have two daughters, Purabi (born in 1934) and Sipra (born in 1938). Both of them are happily married. Purabi works for the Library of Congress in Washington DC and is the head of the section of South Asian languages. Sipra is an Associate Professor of Anthropology at New Paltz College, a part of the State University of New York. Sipra married first. Her husband is Maurice Glen Johnson, an American. He is a Professor of Political Science at Vassar College in Poughkeepsie, New York. They have two children, Denise and Robert. Purabi's husband, Willi W. Schur, is German. He is an engineer and works for the firm of Gibbs and Cox. When Purabi married Willi, my brother-in-law wrote to my wife that now she was an international mother-in-law.

As I have already said, my wife has been my constant companion and helper. After coming to the United States, she learned to drive. Life here is impossible without a car. She takes me to work and brings me home. Also we have traveled by car all over the United States and Europe. In connection with our travels, she has developed her hobby of photography and is now an excellent amateur photographer. We have more than 5000 color slides from all over Europe, the United States and also Egypt, Iran, India, Thailand and Japan. During our travels, we have developed a taste for European food, especially French and Italian. My wife is now an expert cook, making many European dishes, besides Bengali-style Indian dishes. As the universities in which I have worked have had a very active social life, we entertained a lot until I retired; without my wife's help and company I would not be the man I am.

Publications and References

[1] Bose, R. C. (1932) On the number of circles of curvature perfectly enclosing or perfectly enclosed by a closed convex oval. *Math. Z.* **35**, 16–24.

[2] BOSE, R. C. (1936) On the exact distribution and moment-coefficients of the D^2-statistics. *Sankhyā* **2**, 143–154.

[3] BOSE, R. C. (1939) On the construction of balanced incomplete block designs. *Ann. Eugen., London* **9**, 358–399.

[4] BOSE, R. C. (1947) Mathematical theory of the symmetrical factorial design. *Sankhyā* **8**, 107–166.

[5] BOSE, R. C. (1973) *Graphs and Designs*. Edizioni Cremonese, Rome, 1–104. (Based on a course of eight lectures delivered at the CMIE Summer Institute on Finite Geometrical Structures and their Applications, Bressanone, Italy, June, 1972).

[6] BOSE, R. C. and CONNOR, W. S. (1952) Combinatorial properties of group divisible incomplete block designs. *Ann. Math. Statist.* **23**, 357–383.

[7] BOSE, R. C. and KISHEN, K. (1940) On the problem of confounding in the general symmetrical factorial design. *Sankhyā* **5**, 21–36.

[8] BOSE, R. C., MAHALANOBIS, P. C. and ROY, S. N. (1936) Normalization of variates and the use of rectangular coordinates in the theory of sampling distributions. *Sankhyā* **3**, 1–40.

[9] BOSE, R. C. and MESNER, D. M. (1959) On the linear associative algebras corresponding to association schemes of partially balanced designs. *Ann. Math. Statist.* **30**, 21–38.

[10] BOSE, R. C. and NAIR, K. R. (1939) Partially balanced incomplete block designs. *Sankhyā* **4**, 19–38.

[11] BOSE, R. C. and RAY-CHAUDHURI, R. K. (1960) On a class of error detecting binary codes. *Information and Control* **3**, 68–79.

[12] BOSE, R. C. and ROY, S. N. (1938) Distribution of the studentized D^2-statistic. *Sankhyā* **4**, 19–38.

[13] BOSE, R. C. and SHIMAMOTO, T. (1952) Classification and analysis of partially balanced incomplete block designs with two associate classes. *J. Amer. Statist. Assoc.* **47**, 151–184.

[14] BOSE, R. C. and SHRIKHANDE, S. S. (1959) On the falsity of Euler's conjecture about the non-existence of two orthogonal Latin squares of order $4t+2$. *Proc. Nat. Acad. Sci. USA* **45**, 734–737.

[15] BOSE, R. C. and SHRIKHANDE, S. S. (1960) On the construction of sets of mutually orthogonal Latin squares and the falsity of a conjecture of Euler. *Trans. Amer. Math. Soc.* **95**, 191–209.

[16] BOSE, R. C., SHRIKHANDE, S. S. and BHATTACHARYA, K. N. (1953) On the construction of group divisible incomplete block designs. *Ann. Math. Statist.* **24**, 167–195.

[17] BOSE, R. C., SHRIKHANDE, S. S. and PARKER, E. T. (1960) Further results on orthogonal Latin squares and the falsity of Euler's conjecture. *Canad. J. Math.* **12**, 189–203.

[18] BOSE, R. C. and SRIVASTAVA, J. N. (1964) Analysis of irregular factorial fractions. *Sankhyā* A **26**, 117–144.

[19] CHENG, CHING-SHUI (1978) Optimality of certain asymmetrical experimental designs. *Ann. Statist.* **6**, 1239–1261.

[20] CHENG, CHING-SHUI (1981) The comparison of PBIB designs with two associate classes.

[21] EULER, L. (1782) Recherches sur une nouvelle espèce des quarrés magiques. *Fern. Zeevwsch Genoot. Weten. Vliss.* **9**, 85–239.

[22] FISHER, R. A. (1942) The theory of confounding in factorial experiments in relation to the theory of groups. *Ann. Eugen., London* **11**, 341–353.

[23] FISHER, R. A. (1945) A system of confounding for factors with more than two alternatives giving completely orthogonal cubes and higher powers. *Ann. Eugen., London* **12**, 283–290.

[24] HOTELLING, H. (1931) The generalization of the 'Student ratio.' *Ann. Math. Statist.* **2**, 360–378.

[25] KIEFER, J. (1975) Construction and optimality of generalized Youden designs. In *A Survey of Statistical Design and Linear Models*, ed. J. N. Srivastava, North-Holland, Amsterdam, 333–353.

[26] MAHALANOBIS, P. C. (1930) On tests and measures of group divergence Part I: theoretical formulae. *J. Asiatic. Soc. Bengal* **26**, 541–588.

[27] PARKER, E. T. Construction of some sets of mutually orthogonal Latin squares. *Proc. Amer. Math. Soc.* **10**, 964–1951.

[28] SRIVASTAVA, J. N. (ED.) (1973) *A Survey of Combinatorial Theory* (with the cooperation of F. Harary, C. R. Rao, G. C. Rota, and S. S. Shrikhande) North-Holland, Amsterdam.

[29] TARRY, G. (1900) Le problème des 36 officiers. *C. R. Ass. Franç. Av. Sci. Nat.* **1**, 122–123.

Wassily Hoeffding

Wassily Hoeffding was born at Mustamäki, Finland, near Leningrad, on 12 June 1914. After completing his high school education in Berlin, he entered the Handelshochschule in 1933 and later read mathematics at Berlin University. He obtained his Ph.D. in 1940 with a dissertation in correlation theory.

During the war, he worked in Berlin partly as an editorial assistant, and partly as a research assistant in actuarial science. He left Germany for the USA in 1946 and settled first in New York where he attended lectures at Columbia University. In 1947, he was invited to join the Department of Statistics at the University of North Carolina, Chapel Hill, and has remained there ever since. He became a full professor in 1956, Kenan Professor in the University in 1973, and retired in 1979.

Professor Hoeffding has had a variety of research interests in statistics, ranging from correlation theory through U-statistics to asymptotic methods. He is a Fellow of the Institute of Mathematical Statistics and the American Statistical Association; he was elected a member of U.S. National Academy of Sciences in 1976. He was President of the Institute of Mathematical Statistics in 1969.

Professor Hoeffding enjoys nature, travel, and discussions with good friends. He remains vitally interested in "the lure of the subject."

A Statistician's Progress from Berlin to Chapel Hill

Wassily Hoeffding

1. Childhood and Education

I was born in 1914 in Mustamäki, Finland, near St. Petersburg (now Leningrad). Finland was at that time part of the Russian Empire. My father, whose parents were Danish, was an economist and a disciple of Peter Struve, the Russian social scientist and public figure. An uncle of my father's was Harald Hoeffding, the philosopher. My mother, née Wedensky, had studied medicine. Both grandfathers had been engineers.

In 1920 we left Russia for Denmark, where I had my first schooling. In 1924 the family settled in Berlin. In high school, an Oberrealschule which put emphasis on natural sciences and modern languages, I liked mathematics and biology and disliked physics. When I finished high school in 1933, I had no definite career in mind; there was no equivalent of the four-year college in Germany. I thought I would become an economist like my father and entered the Handelshochschule (later called Wirtschaftshochschule) in Berlin. But I soon found that economics was too vague a science for me. Chance phenomena and their laws captured my interest. I performed series of random tossings and recorded their outcomes before I knew much about probability theory. One of the few books on chance phenomena that I found in the library of the Hochschule was *Die Analyse des Zufalls* by H. E. Timerding, and it fascinated me. In 1934 I entered Berlin University to study mathematics.

Probability and statistics were very poorly represented in Berlin at that time. Hitler had become Chancellor of the Reich in 1933 and Richard von Mises, who was Jewish, had already left the university. There was one course in mathematical statistics, taught by Alfred Klose, a disciple of von Mises. He was also a Nazi and evidently felt it his duty to emancipate himself from the influence of his Jewish teacher. Still, von Mises' *Wahrscheinlichkeitsrechnung* was the textbook he used, and he followed it closely.

Advanced calculus was taught by Erhard Schmidt (of Hilbert–Schmidt expansion fame). Plump, bald, with a finely chiseled, clean-shaven square face, speaking with a pronounced Baltic-German accent, Schmidt was an excellent classroom teacher. In those years in Germany a person's attitude to Nazism was all-important. Of course, nobody would publicly declare his anti-Nazi feelings. But just by observing a person, by noticing what he did and did not say, one could sense where he stood. Thus it soon became clear to me that Erhard Schmidt, although a German patriot, could not possibly approve of Nazism.

Unfortunately, being a fine mathematician did not make a person immune to the Nazi infection. An outstanding example was Ludwig Bieberbach, with whom I took several courses. He was the founder and editor of the journal *Deutsche Mathematik*. I once went to a public lecture he gave on "Aryan" and "non-Aryan" mathematics. A distinguishing mark of "Aryan" mathematicians, according to him, was that they liked to appeal to geometric intuition; this was certainly present in Bieberbach's work. After the lecture he invited his listeners, mostly students, to hand him written questions. One questioner embarrassed him by asking him to explain why the non-Aryan Richard Courant so often used geometric arguments in his books.

In my first semester at the university I took a course in elementary number theory with Alfred Brauer (a brother of Richard Brauer). He was unquestionably the best teacher I ever had, but he soon had to leave the university. When I came to the University of North Carolina in 1947, it was a pleasant surprise to find that Alfred Brauer was teaching there.

The meager fare in mathematical statistics that I was fed in my lectures in Berlin, I tried to supplement by reading journals. But somehow I did not fully absorb the spirit of research at the frontier of the subject in my student days. My Ph.D. dissertation [6] was in descriptive statistics and did not deal with sampling. It was concerned with properties of bivariate distributions that are invariant under arbitrary monotone transformations of the margins. It thus touched on rank correlation, some of whose sampling aspects I later explored [7], [8].

My *Doktorvater* or Ph.D. supervisor was Klose. I chose the topic of the thesis and worked on it largely by myself, with some suggestions and encouragement from him. He was a Baltic German and had his own ideas about Russians. He warned me to refrain from making exaggerated claims in my thesis that I could not substantiate as, he thought, Russians were prone to do.

2. Earning a Living in Wartime Germany

On completing my studies in 1940, I accepted two part-time jobs: as an editorial assistant with the *Jahrbuch über die Fortschritte der Mathematik* and as a research assistant with the inter-university institute for actuarial science (Berliner Hochschulinstitut für Versicherungswissenschaft). I held both jobs until almost the end of the war. I never applied for a teaching position in Germany: I had been stateless since leaving Russia and did not wish to acquire German citizenship, which was necessary to hold a university teaching job.

The actuarial institute had just been established. I was charged with building up its library, but since few books and journals could be bought during the war, this job remained largely a sinecure.

The editor of the *Jahrbuch* was Harald Geppert. He simultaneously edited the *Zentralblatt für Mathematik* from the same office. Practically all the current mathematical literature that entered wartime Germany from abroad must have reached our two office rooms.

My colleagues at the *Jahrbuch* represented an interesting cross-section of German mathematicians during the war. Harald Geppert, a differential geometer of wide mathematical interests, half-Italian, although a Party member, remained an honorable man. In 1944, stateless persons "of German or related blood" were declared to be subject to military service. Having a Danish name, I was adjudged to fall in this category. My diabetes saved me from serving in the German army, but the threat of having to do labor service for the German war effort seemed very real. At that time Geppert suggested that I do some mathematical work with military applications. Knowing that I could be frank with him, I said that doing this kind of work would be contrary to my conscience. The conversation took place at his home, at night. Also present was Hermann Schmid, Geppert's assistant on the *Zentralblatt* who became its editor after the war. Schmid was the scion of a Prussian military family, a very reserved person, and I felt that I could trust his sense of honor. After the war I learned that when the Soviet army was about to enter Berlin, Geppert, at the breakfast table, gave poison to his small son, and then took poison with his wife.

The oldest person on the *Jahrbuch* staff was Max Zacharias, a little man with a goatee, a retired school teacher, author of the article on elementary geometry in the *Enzyklopädie der mathematischen Wissenschaften*. He was a fervent Nazi. When the outcome of the war was already clear, he told his colleagues in the office about a secret weapon being prepared that was sure to decide the war in Germany's favor.

Fräulein Doktor S., a tall blonde young lady, was a true believer, without guile and without any apparent doubts in Hitler. Fritz D., in contrast, a young algebraist who did work on coding, never discussed politics in the office. This fact alone betrayed his position.

Towards the end of the war some French and Dutch prisoners of war were released to work on the *Jahrbuch* staff. One of them, a Dutchman from Frisia, confided to me his deep hatred for Hitler. After the war I learned that some of these prisoners were harshly criticized at home for working for the Germans.

In February 1945 I left Berlin with my mother for a small town in the province of Hanover to stay with a Swiss friend of my father's. Klose made this an official transfer. My father stayed behind and was captured by what was later to become the KGB. He had been employed for many years at the office of the American Commercial Attaché and then had been the economic correspondent of American and Swiss periodicals. This made him a "spy" in the eyes of the KGB.

Hanover soon became part of the British zone of occupation. My mother and I stayed there for over a year, vainly trying to help release my father. My younger brother Oleg, who had spent the war in London working in the Economic Warfare Division of the U.S. Embassy, visited us in American uniform. I asked him to send me a copy of the recently published Volume I of M. G. Kendall's *The Advanced Theory of Statistics*. It read to me like a revelation. It was in Hanover that I wrote my first statistical paper in the modern sense of the word [7]. It established the asymptotic normality of Kendall's rank correlation coefficient τ in the general case of independent identically distributed random vectors. It so happened that the paper was published in the same issue of *Biometrika* as the proof by H. E. Daniels and M. G. Kendall [4] of the analogous result for the case of sampling from a finite population.

3. Settling in America

Having lost all trace of my father, we left Germany for Switzerland and arrived in New York City in September 1946. (My father later escaped from his prison in Potsdam.) As I was unemployed, I attended lectures at Columbia University by Abraham Wald, Jack Wolfowitz, and Jerzy Neyman, who was then visiting Columbia. I was in the thick of contemporary statistics. I remember how Neyman questioned me on the effects of the allied bombing of Berlin, which he had tried to estimate during the war.

In 1940 I had sent a few copies of my Ph.D. thesis to statisticians in other countries, including the United States. These, and my one personal copy, are the only ones that survived the war. The rest, stowed in three different locations in Berlin, were all destroyed in air raids. Thus my name was not entirely unknown in the USA. My brother Oleg, whose arrival in New York preceded mine by three months and who now was an economics instructor at Columbia, helped me to get invitations from the Cowles Commission for Economic Research (then at Chicago University) and from Harold Hotelling, who had just established the Department of Mathematical Statistics at the University of North Carolina in Chapel Hill.

I first went to Chicago to give a talk on what I later called *U*-statistics. The standard length of a mathematical seminar at Berlin University was $1\frac{1}{2}$

hours, and it did not occur to me that the length could be different at Chicago. As I was speaking on and on, I began to notice some restlessness around me, until I finally took the polite hint from the chairman. Still, my hosts, including Jacob Marshak and Tjalling Koopmans, were most gracious. I went to see Paul Halmos, whose paper on unbiased estimation [5], which was related to the topic of my talk, had just appeared.

Soon after, a letter from Hotelling offered me a position as research associate at his new department. He did not ask for a preliminary visit; later he said he had been impressed by the fact that a Ph.D. thesis in mathematical statistics had come out of Germany. Hotelling's offer was more congenial to my interests than the one from Chicago. In May 1947 I arrived in Chapel Hill.

4. Statistics at the University of North Carolina at Chapel Hill

When I met Harold Hotelling in Chapel Hill, the crest of his scientific activity was behind him. While he was yet to publish a number of respectable research papers, what I witnessed was the afterglow of a great mind. Perhaps the one human trait that best characterized him was the goodness of his heart. He was always ready to help those around him, and this in many ways. Apart from launching me on my academic career, he volunteered (to give a single example from my own experience) to help finance the purchase of my first car.

Hotelling's department had been started a year before I came. By now he had assembled a small group of outstanding people. I found much in common with Herbert Robbins. A postcard from his summer retreat in Vermont resulted in a joint paper on the central limit theorem for sums of m-dependent random variables [11]. (In a book on non-parametric statistics it has been stated that the theorem was developed with a view to non-parametric applications; this is not so.) I sat in on some of the finely polished courses of P. L. Hsu, who was then already in frail health. In 1948, when the victory of the Chinese revolution was approaching, he felt it his patriotic duty to return to his country. Later that year, the first course that I gave, which was in multivariate analysis, was based on the notes I had taken in Hsu's lectures. Among the statistics students I had met in Chapel Hill was George Nicholson, who soon joined the faculty of the department and later became its chairman.

Around the time when I arrived in America, I found another proof of the asymptotic normality of τ which turned out to be applicable to a large class of statistics. This led to my paper on U-statistics [8]; incidentally, the "U" stems from "unbiased estimator." I like to think of this paper as my "real" Ph.D. dissertation.

Quite a few papers on U-statistics have appeared since then. The definitive form of the Berry–Esseen type bound for non-degenerate U-statistics has been found only quite recently by Borovskikh [1].

I was to remain in Chapel Hill until my retirement and beyond. Congenial colleagues, a relaxed, informal academic life style, the attractive nature of the town, the relative closeness of the sea and the mountains, combined with an inborn inertia, made me resist the temptations of moving to other campuses. Being somewhat reserved by nature, I cherish all the more the friendships and contacts I have had with my colleagues and students in the department and their families.

Of the many visitors who came to teach and do research in the department, a few have specially impressed themselves on my memory. E. J. G. Pitman taught in Chapel Hill in 1948–49 in his masterful manner. His knack in seeking out beauty in statistics is again reflected in his recent book [15].

V. V. Petrov of Leningrad University visited Chapel Hill for several weeks in the fall of 1963. He talked to us on aspects of sums of independent random variables, later the subject of his excellent monograph [14].

Shortly before he came, a law had been enacted in North Carolina forbidding communists to speak on campuses. It seemed to extend to any kind of public speech, including mathematical lectures. We had no idea whether Petrov was a Communist Party member and did not want to ask him. George Nicholson, who was then chairman of the department, found a brilliant solution to the problem. He called an appropriate government agency in Washington (the FBI, I believe) and asked whether they knew if Petrov was a member of the Soviet Communist Party. Of course, they had no such information. Nicholson determined that this made it legal for Petrov to speak in our seminar. Petrov was embarrassed to find himself made a center of attention in the local press.

In 1973–74 we had the good fortunate to have Ildar Ibragimov of Leningrad visit us for several months. He lectured on his joint work with Has'minskii on asymptotic estimation theory (material which was later included in their book [13] and on other topics reflecting his far-flung interests. Both professionally and personally his visit was a great success.

5. Travels and Visits

There were many opportunities to go to scientific gatherings—occasions for meeting new people, renewing old acquaintances, and seeing the world.

The Second Berkeley Symposium, held in 1950, was the first one I attended. Apart from seeing Jerzy Neyman again, I met statisticians and probabilists from the West Coast and from all over the world.

After the symposium, Sudhish Ghurye, Gopinath Kallianpur (then students at Chapel Hill), Miriam Yevick and I went on a three-day hike in the

Yosemite Park. It remains one of the most delightful memories of my life. I was very pleased when Kallianpur became professor at Chapel Hill after I retired in 1979.

In the fall of 1955, following an invitation from Herbert Robbins, I spent a semester at Columbia University. William Feller was then commuting from Princeton to lecture on probability theory. Before the lectures, in the office he shared with me, he spent a few minutes in visible intense concentration. His lectures were brilliant but often difficult; some students complained that he was a bad lecturer.

At that time I had obtained some curious results about the distribution of the maxima of the consecutive partial sums of independent, identically distributed random variables. When I showed them to Feller, he found them interesting and thought them to be new; they reminded him of a thesis he had read not long before, by E. Sparre Andersen. It turned out that Sparre Andersen's paper [17] had by now been published, and I found in it not only my results but more.

The summers of 1956 and 1957 were spent at Cornell University. Jack Wolfowitz and I wrote a joint paper on the distinguishability of families of distributions [12]. Jack, Lionel Weiss and I went on walks in the hills around Ithaca during which we would discuss Israel and mathematics. It was then in Ithaca that I first met Jack Kiefer.

In 1962 I attended the International Congress of Mathematicians in Stockholm. I was especially anxious to meet Kolmogorov. Before going there, I wrote a short paper [9] for the *Theory of Probability and its Applications*, and sent it to Kolmogorov, who was the editor-in-chief. At the Congress I received a note from Professor Frostman, one of the organizers of the Congress, asking me to chair a session at which Kolmogorov was to present a paper by Dynkin, who had been prevented from coming. Before the session I approached Kolmogorov and asked about my paper: "Yes, yes, your paper has been accepted," he said, and that ended the conversation. When I was standing on the podium, ready to introduce the speaker, I suddenly found Professor Frostman making the introduction. It was evident that Kolmogorov, for some reason, did not want to be introduced by me.

Harald Cramér, who was then the rector of Stockholm University, was the host at a reception in the Ghost House and told us about its history. I was to meet him and his wife Marta again frequently, when he paid prolonged visits to Chapel Hill and the Research Triangle Institute, working on his joint book with Ross Leadbetter [3]. In Harald Cramér, a penetrating scientific mind is happily welded with a warmly human personality.

Among those whose acquaintance I made in Stockholm was Yuri Linnik. His keen analytic powers and his prodigious energy were then being increasingly directed toward probability and statistics. He had a fine sense of humor. When, two years later, he came to the Indian Statistical Institute in Calcutta where I was visiting, I congratulated him on his election to the Soviet Academy of Sciences. "Oh, the only difference this makes for me is that I will receive a higher salary," he replied. Later that year, when he met me at the Leningrad airport, he immediately started talking mathematics. When he was organizing the session on non-parametric statistics to be held at the meeting of the International Statistical Institute in Sydney in 1967, he asked me to take part. In Sydney I reminded him that now we had met on four continents. "Surely, next time we shall meet in Africa," he said. But to my regret, I never saw him again; he died in 1972.

My paper on asymptotically optimal tests for multinomial distributions [10] took as its starting point Sanov's [16] results on probabilities of large deviations. Its findings have been extended in interesting ways by several authors. The paper by Lawrence D. Brown [2] is particularly remarkable.

In 1964–65 I spent six months in India. It was an appointment to the Research and Training School of the Indian Statistical Institute in Calcutta, arranged by C. R. Rao and sponsored by UNESCO. I became acquainted with the highly-reputed Institute and the glaring contrasts of Calcutta life. My stay in India was capped by a one-month lecture tour of the country, also made possible by the untiring efforts of C. R. Rao. I visited Benares, Lucknow, Delhi, Agra, Bombay, Bangalore, Mysore City, Trivandrum, Madras. I was pleased to meet in a number of these cities old friends whom I knew from Chapel Hill as students or visitors.

6. An Exchange Visit to Russia

After India I was to go on a one-month visit to Russia under the inter-academy exchange agreement. The U.S. National Academy of Sciences was at first reluctant to sponsor my trip because they thought that I had little chance of obtaining a Soviet visa. When I pointed out that two years earlier I had been in Russia as a tourist, they agreed to support the trip. In Calcutta the Soviet visa was handed to me on the morning of the day I was to depart for Delhi and Moscow. Later I learned that this was, and apparently still is, a common experience of exchange visitors to Russia.

The trip took place in April. I visited Moscow, Leningrad, Kiev, Tashkent and Novosibirsk. Tashkent was included at the insistance of S. H. Sirazhdinov, who, with Yuri Prohorov, had visited the Indian Statistical Institute during my stay there. I was cordially received by old friends and new. Later I was asked whether, while in Russia, I was followed in the streets or whether my luggage had been tampered with. I don't know: I never checked.

In Moscow, Prohorov and Zolotarev suggested that I accompany them on a short trip to Vilnius to get acquainted with the active group of probabilists working there. But at the American embassy I was discouraged from going to Vilnius on the ground that the United States had not recognized the incorporation of the Baltic countries into the Soviet Union. I found this policy unwise but complied with the embassy's request. Ten years later, after the meeting of the International Statistical Institute in Warsaw in 1975, I spent one day in Vilnius and was cordially welcomed there by V. A. Statulevičius and his colleagues.

The visit to Akademgorodok, the seat of the Siberian section of the Academy of Sciences of the USSR, near Novosibirsk, was of special interest. Few Westerners had been there before me. I was warmly received by A. A. Borovkov and his co-workers, mostly younger people. I had arrived directly from Tashkent, where flowers were in bloom in the public squares. Here, when we were walking on the ice-bound Ob River in a chilly wind, I put on the only head covering I had with me, an embroidered Uzbek scull cap which had been presented to me in Tashkent.

7. Concluding Remarks

In April 1979, the year I was to reach the age of 65 and to retire from teaching, a symposium on the asymptotic theory of statistical tests and estimation was held in Chapel Hill. It was organized by the efforts of Indra Chakravarti. For me this was a welcome occasion to greet old friends and to meet new ones, some of whom I knew by correspondence. Unfortunately, in the middle of the symposium banquet, I had to leave for the hospital, where my right leg had to be amputated. (The reason was an infection related to my diabetes.) Since then I have been getting used to a new kind of life.

Ever since I switched from economics to probability and statistics in my early student days, this area has continued to absorb my interests. The very idea that the seeming chaos of chance obeys mathematical laws is immensely attractive. It gives me great satisfaction to have made a few contributions to the understanding of this field.

The successes I had did not come easy to me. They were the fruit of long hours of work which often led to dead ends. I am well aware that with advancing years my capacity to work has diminished. The lure of the subject persists. Whether I will contribute more to it, only time will tell.

Publications and References

[1] Borovskikh, Yu V. (1979) Approximation of U-statistics distribution (in Russian). *Dokl. Akad. Nauk Ukrain. SSR* **9**, 695–698.

[2] Brown, L. D. (1971) Non-local asymptotic optimality of appropriate likelihood ratio tests. *Ann. Math. Statist.* **42**, 1206–1240.

[3] Cramér, H. and Leadbetter, M. R. (1967) *Stationary and Related Stochastic Processes*. Wiley, New York.

[4] Daniels, H. E. and Kendall, M. G. (1947) The significance of rank correlations where parental correlation exists. *Biometrika* **34**, 197–208.

[5] Halmos, P. R. (1946) The theory of unbiased estimation. *Ann. Math. Statist.* **17**, 34–43.

[6] HÖFFDING, W. (1940) Maszstabinvariante Korrelationstheorie. *Schriften des Math. Inst. und des Inst. für angewandte Math. der Univ. Berlin* **5** (3), 181–233.

[7] HÖFFDING, W. (1947) On the distribution of the rank correlation coefficient τ when the variates are not independent. *Biometrika* **34**, 184–196.

[8] HOEFFDING, W. (1948) A class of statistics with asymptotically normal distribution. *Ann. Math. Statist.* **19**, 293–325.

[9] HOEFFDING, W. (1964) On a theorem of V. M. Zolotarev (in Russian). *Teor. Verojatnost. i Primenen.* **9**, 96–99. (English translation: *Theory Prob. Appl.* **9**, 89–91.)

[10] HOEFFDING, W. (1965) Asymptotically optimal tests for multinominal distributions. *Ann. Math. Statist.* **36**, 369–401.

[11] HOEFFDING, W. and ROBBINS, H. (1948) The central limit theorem for dependent random variables. *Duke Math. J.* **15**, 773–780.

[12] HOEFFDING, W. and WOLFOWITZ, J. (1958) Distinguishability of sets of distributions. *Ann. Math. Statist.* **29**, 700–718.

[13] IBRAGIMOV, I. A. and HAS'MINSKII, R. Z. (1979) *Asymptotic Theory of Estimation* (in Russian). Nauka, Moscow.

[14] PETROV, V. V. (1972) *Sums of Independent Random Variables* (in Russian). Nauka, Moscow. (English translation: Springer, New York, 1975).

[15] PITMAN, E. J. G. (1979) *Some Basic Theory for Statistical Inference*. Chapman and Hall, London.

[16] SANOV, I. N. (1957) On the probability of large deviations of random variables (in Russian). *Mat. Sb.* N. S. **42** (**84**), 11–44. (English translation: *Select. Transl. Math. Statist. Prob.* **1** (1961), 213–244.)

[17] SPARRE ANDERSEN, E. (1953) On the fluctuations of sums of random variables. *Math. Scand.* **1**, 263–285.

E. J. G. Pitman

Edwin James George Pitman was born in Melbourne on 29 October 1897, of English parents. He was educated at South Melbourne College, and later read mathematics and physics at the University of Melbourne. He graduated with a BA in 1921, after serving with the 14th Battalion of the AIF in 1918–19.

He was appointed Acting Professor of Mathematics at Canterbury College, University of New Zealand in 1922, and returned to Melbourne in 1924 to become Tutor in Mathematics and Physics at Ormond and Trinity Colleges, and part-time lecturer in physics at the University. He became Professor of Mathematics at the University of Tasmania in 1926, a position which he held for 37 years before retiring at the end of 1962.

Professor Pitman travelled to the USA for the first time as a visiting professor in 1948–49, when he lectured at Columbia, North Carolina and Princeton. He returned to Stanford in 1957, Johns Hopkins in 1963, and Chicago in 1968–69. He was a visiting Senior Research Fellow at the University of Dundee, Scotland, in 1973.

He has been closely involved in the development of mathematics and statistics in Australia. He was President of the Australian Mathematical Society in 1958–59 and became an honorary Life Member in 1968. He was made an honorary Life Member of the Statistical Society of Australia in 1966. The Statistical Society's highest honour, the gold "Pitman Medal" was named after him; he was presented with the first medal in 1978.

Among his other honours are Fellowship of the Australian Academy of Science in 1954, membership of the International Statistical Institute in 1956, and Honorary Fellowship of the Royal Statistical Society in 1965. He was also awarded an honorary D.Sc. by the University of Tasmania in 1977.

Professor Pitman has written a book and several papers on a variety of topics in probability and statistics. His recreational interests are gardening, music, and drama. He married Elinor Hurst in 1932. They have four children, Jane (Reader in Mathematics, University of Adelaide), Mary (Mrs Baldwin, Associate Professor of Chemistry, and Assistant Dean of Students, Concordia University, Montreal) Edwin (Civil Engineer, Department of Main Roads, Tasmania), and James (Associate Professor of Statistics, University of California, Berkeley).

Reminiscences of a Mathematician Who Strayed into Statistics

E. J. G. Pitman

1. My Introduction to Statistics

I remember very clearly my first encounter with statistics, at the age of 23. In a course called Advanced Logic at the University of Melbourne, Professor W. R. Boyce-Gibson, a very able philosopher and an excellent lecturer, devoted two or three lectures to the subject. I decided then and there that statistics was the sort of thing that I was not interested in, and would never have to bother about.

Four years later, in 1925, I applied for the Professorship of Mathematics at the University of Tasmania. Applicants were asked to state whether they had any knowledge of statistics, and if they would be prepared to teach a course in the subject. I wanted the appointment, so in my application I wrote "I cannot claim to have any special knowledge of the Theory of Statistics; but, if appointed, I would be prepared to lecture on this subject in 1927." I think that the word *special* could have been deleted with accuracy; but I was being careful not to exaggerate.

My application was successful. In 1926 I was appointed Professor of Mathematics in the University of Tasmania, and so I was committed to teaching statistics, and therefore, in spite of my youthful aversion to the subject, committed to studying it. It happened that I was also very soon involved in applying it. In fact, I began almost simultaneously the three activities of learning, teaching, and applying statistical theory. In 1929 I gave a third-year course, Theory of Statistics.

I learnt much of the basic theory from G. Darmois's excellent *Statistique Mathématique*, published in 1928, which, according to the 14-page critically commending preface by M. Huber, Directeur de la Statistique Générale de la France, was the first important French work on modern statistical theory. Borel's many-volumed treatise on probability was being published, and that was a great help.

Some time about 1928, a part-time student in mathematics, R. A. Scott, who worked in the State Department of Agriculture, brought me one day the data and statistical analysis of some NPK (nitrogen, phosphorus and potassium) field trials on potatoes, which he asked me to comment on before he sent them for final criticism to Dr John Wishart in Cambridge. He also produced, open at the appropriate page, Fisher's *Statistical Methods for Research Workers*, which I had not seen or heard of. I took the papers and the book home. After studying them that night, I decided by the light of common sense that the calculations were sensible and according to Fisher, and reported next morning that, as far as I could see, everything was correct. This was confirmed later by Wishart. That was my introduction to Fisher, and, in particular, to the analysis of variance. It was also the beginning of a collaboration in field trials with Scott and the Department of Agriculture.

I very soon began to wonder about the validity of applying normal variable theory to the randomised blocks and Latin squares of our field trials. One can go a long way in the analysis of variance without invoking the normal genie. It is only in determining significance levels and power that distribution theory comes in. These wonderings led me later into non-parametric methods; but an immediate result was the invention of the semi-Latin square, an arrangement of $2n$ treatments applied to $2n^2$ plots arranged in $2n$ rows and n columns. Others may have used the design before or since; but I am quite sure that inadvertently I named it. I remember saying to Scott when we were planning experiments "This is a design I should like to try. It is a sort of semi-Latin square." A generalisation, which is what we actually used, is mn treatments applied to mn^2 plots arranged in mn rows and n columns. In the published results [14] Scott called our arrangement with $n=4$, $m=5$, a semi-Latin square. A better name would be partial Latin square.

I had two reasons for evolving it. Firstly, our Latin squares in the field were not squares but rectangles. This was because the plots were not squares, but long rectangles, rows of potatoes. I thought that the effect of variation in fertility of the soil would be reduced if the whole area was more nearly square.

Secondly, it was not a standard design; but with the usual model with independent residuals with zero means and a common variance, the ordinary analysis was valid. I wanted to get the reaction of the statistical establishment, in order to find out what model they really used. The design was first approved by Fisher, who said that Student had used it, and by Wishart. Later [15], it was condemned by Yates and Fisher on the ground that with the usual analysis it gave a biased estimate of error. They were then using a permutation model. My suspicions were confirmed.

2. Early Publications

I started to study probability and statistics in 1926, but I did not publish anything until 1936. There were two main reasons for the delay in publication; the work load I had to carry, and the nature of my upbringing.

When I went there in 1926, the University of Tasmania was a very small university which was just starting to grow. The Department of Mathematics consisted of myself and a part-time lecturer. By 1948 it had grown to a professor, an associate professor, two lecturers, and two tutors. I had to teach a wide range of mathematics, pure and applied. For many years I gave 12 lectures a week, and we worked on Saturday mornings. Let me say that at the time I did not think this abnormal.

As a student I had three years of university mathematics, interrupted by two years in the army during the First World War, and followed by a year of mainly physics. In those days there were no graduate schools in mathematics in Australian universities. Some of the other universities had scholarships which enabled promising young mathematicians to go abroad for further study, mostly to Cambridge, England.

When I left the University of Melbourne after four years of study there, I had had no training in research; but I thought that I had learned to study and use mathematics, and that I would be willing to tackle any problem that arose, even if that necessitated learning a new branch of mathematics. I regarded myself as a general practitioner with most of my learning still to be done, and not as a specialist.

The lack of research training was a handicap which took me some time to overcome. One result was that for many years I never gave a thought to printed publication. Of course, much of my original work went into my lectures on various branches of mathematics. What eventually spurred me to publish was a government research grant to the University. In 1936 the Commonwealth Government decided to provide £30,000 per year for a period of five years for the promotion of scientific research in the universities. On a state population basis, Tasmania's share of this would have been 1/30; but we were allocated more than double this, £2,400.

I was chairman of the Professorial Board at the time, and had taken part in the preliminary negotiations. I was determined that when we presented our first annual report, we should have more than progress and promises to write about. The only satisfactory report is published work. I immediately set to work to write my first paper for publication. My first try was on a topic in hydrodynamics. When I had typed it out, I decided that it was not worth publishing, so I had to look for something else. I had already obtained the general form of distributions admitting sufficient statistics. With this as a start, I wrote a paper entitled "Sufficient Statistics and Intrinsic Accuracy." [4].

I sent the paper to Dr John Wishart of Cambridge. I had had no direct contact with him; but I thought he would know my name from my connection with the Department of Agriculture's field trials, the results of which had always been sent to him for comment. He reacted magnificently. He wrote back warmly welcoming me as a mathematician interested in statistics, and said that the paper would be published in the *Proceedings of the Cambridge Philosophical Society*. Not only did he arrange for the publication of that paper and the seven other papers which I wrote in the next two years, but he also did all the proof-correcting of the eight papers.

I met Wishart only once, in 1949, at dinner at Sam Wilks's house in Princeton. It gave me the opportunity to express in person, what until then I had been able to do only by letter, my great gratitude for his marvelous kindness in helping me at a critical stage. I was surprised to find that he was a voluble talker. Soon after dinner, Wilks's son turned on the television set in a distant corner of the large room in which we were sitting. Television was then fairly new. Gradually the guests drifted over to the television set, and conversation ceased. Most of the programme was an old Charlie Chaplin film, which was obliterated when from time to time an aircraft flew overhead. At the end of the evening I left with Wishart and somebody else. When we got outside, Wishart exploded. He had been bottled up most of the evening, and felt he had been suppressed. That was the last I saw of him; he was drowned later that year.

It was lucky for me that by the time I started to write, airmail services had started. The combination of that and Wishart's assistance made publication very quick, a most encouraging thing for a beginner. One of my papers, the second I think, was carried in an aircraft which came down in the sea. The mail was salvaged, and my paper arrived at Cambridge wet through, but still legible after drying.

3. Non-parametric Methods and Estimation

When I started to study it, statistical theory was dominated by the normal distribution. Most of the statistical tests in common use were based on the assumption that the distributions sampled were normal, and that worried me. For example, the t test for the significance of difference of means of two samples assumes that the distributions are normal with the same variance. We can devise a corresponding test for samples from two continuous distributions of any given form which differ only in location (see [5]); but what can we do without making any assumptions about the specific form of the distributions?

Suppose we have samples

$$x_1, \dots, x_m \text{ mean } \bar{x};\ y_1, \dots, y_n \text{ mean } \bar{y},$$

and wish to test the hypothesis that the distributions sampled are the same, against the alternative that they differ only in location. If the two distributions are the same, then, instead of taking separate samples of sizes m and n, we might, without affecting any probabilities, first take a sample of $m+n$, and then separate it into a sample of m and a sample of n, by a process which makes all such separations equally probable. We then compare the value of $|\bar{x}-\bar{y}|$ with the values corresponding to all the $\binom{m+n}{m}$

separations. If the distributions differ in location, the observed value of $|\bar{x} - \bar{y}|$ will be likely to be one of the high ones. The rest is well known. I am merely trying to explain the train of thought which started me off writing three papers [5], [6], [7] on distribution-free tests, non-parametric tests as they came to be called.

I called this test the *spread* test. Except when the sample sizes are very small, the labour of making it can be very great. I did not think of very greatly reducing it by replacing the $m + n$ observed numbers by their ranks. Had I done so, I should have had the Wilcoxon test in 1937, for that is what the spread test becomes when the ranks are used in place of the actual numbers. I did investigate an approximation to the distribution of $(\bar{x} - \bar{y})^2$ over the different separations. The use of this turned out to be exactly equivalent to using the ordinary t test. Having kicked the normal distribution out of the door, I was surprised to see it fly in at the window.

There was some difficulty about getting "The Estimation of the Location and Scale Parameters of a Continuous Population of Any Given Form" published. Wishart first sent it to *Biometrika*; but it was refused on the grounds that it was too mathematical for that journal. He next tried the London Mathematical Society, and was again unsuccessful. The editor thought that the paper was too controversial, and he did not want his journal to be cluttered up with correspondence about it. Sir Harold Jeffreys was then asked to sponsor it for the Royal Society; but he refused, saying that it was somewhat like a paper of Neyman's which he had recently been persuaded to sponsor, and which, one gathered, he was not enthusiastic about. Finally it was saved from oblivion by the editor of *Biometrika*, who generously offered to publish it, since no one else would.

All this must seem very strange to present-day statisticians. There is nothing controversial in the paper. It is all mathematics, which no one has ever faulted. It is written in perhaps a somewhat pedestrian style, mainly because I avoided the use of conditional distributions. This was partly because of doubts about my own knowledge of them at that time, but mainly because I feared that some of my readers might not fully understand an argument using conditional distributions, and so might doubt the validity of some of the conclusions. I wanted to be clear and uncontroversial and convincing.

In this paper I introduced the term *estimator* into statistics (see [2]), as a name for the function of the observations, whose observed value is taken as an *estimate* of an unknown parameter. I realised that there was no hope of getting useful, general results about estimators unless the class of estimators under discussion was suitably restricted. "All estimators of a parameter" is the same as "all functions of the observations," and there is always, lurking in the background, the constant function, which gives the correct estimate at one and only one value of the unknown parameter, and is certainly the best estimator at that point.

Let X be a random variable with a distribution having density $f(x-a)$ relative to Lebesgue measure on the real line, the function f being known, but the value of the location parameter a being unknown. What practical, common-sense considerations should restrict the class of a-estimators to be discussed? It seemed to me that if we have no source of knowledge of the value of a, other than a sample of n values of X, any method which would assign the value a_0 to a when the observed values of X were

$$x_1, x_2, \ldots, x_n,$$

would assign the value $a_0+\lambda$ to a when the observed values were

$$x_1+\lambda, x_2+\lambda, \ldots, x_n+\lambda.$$

An estimator A must therefore satisfy the relation

$$A(x_1+\lambda, \ldots, x_n+\lambda) = A(x_1, \ldots, x_n)+\lambda.$$

Similar considerations apply to an estimator C of a scale parameter c when the density is $c^{-1}f(x/c), c>0$. We must have

$$C>0, C(\lambda x_1, \ldots, \lambda x_n) = \lambda C(x_1, \ldots, x_n) \text{ for } \lambda>0.$$

When the density is $c^{-1}f[(x-a)c^{-1}]$, and both a and c are unknown, we require

$$A\left(\frac{x_1+\lambda}{\mu}, \ldots, \frac{x_n+\lambda}{\mu}\right) = \frac{A(x_1, \ldots, x_n)+\lambda}{\mu}, \mu>0$$

$$C>0, C\left(\frac{x_1+\lambda}{\mu}, \ldots, \frac{x_n+\lambda}{\mu}\right) = \frac{C(x_1, \ldots, x_n)}{\mu}, \mu>0.$$

When this is so,

$$\frac{A(x_1, \ldots, x_n)-a}{c}, = A\left(\frac{x_1-a}{c}, \ldots, \frac{x_n-a}{c}\right),$$

has a distribution which is independent of a and c. It is determined by f, which is supposed to be known. The same is true of C/c and $(A-a)/C$.

In this paper only estimators with these properties are considered, and they are thoroughly investigated. I could not think of a suitable adjective to attach to them. They are now called *invariant* estimators. Does anyone in practice use any others?

The doubts about the paper arose, I think, from the presence of the word *fiducial*. I use the term *fiducial function* as a name for a probability density function which appears in the analysis and which is closely related to the likelihood function. The corresponding probability distribution I called the *fiducial distribution*. These are merely technical conveniences: it was necessary to give names to these two things. I do not speak about fiducial probability except to say, in one place, that Fisher does: "As R. A. Fisher expresses it, the fiducial probability of the variable statement (4) is α."

The fiducial distribution in the (a, c)-plane determined by the fiducial function is *not* the probability distribution of the unknown (a, c) determined by the observations. It is merely a distribution from which confidence regions for (a, c) can be determined. I have explained this fully in [9], and shown where I think that Fisher went wrong in his treatment of fiducial probability.

4. The Second World War and My First Visit to the USA

The last of these eight papers was published in the July 1939 issue of *Biometrika*. The Second World War started in September of that year. That put a stop to my flow of papers.

At the beginning of the war, the mathematics staff of the University of Tasmania consisted of myself and Dr J. C. Jaeger, a very able applied mathematician. He was very soon seconded to the CSIRO, the Commonwealth Scientific and Industrial Research Organization for war work. I had to do the best I could with a part-time assistant lecturer. Besides teaching mathematics, I had heavy administrative duties, as I was the Chairman of the Professorial Board, and a member of Council, the governing body of the University. In addition to this, I took on the job (which was really a full-time one) of honorary Education Officer for the Royal Australian Air Force in Tasmania, in charge of selection and pre-enlistment training of air crew, and responsible for determining and accepting or not, the educational standards of all Tasmanian applicants for entry into the Air Force. Later, when the Air Training Corps (for youths under enlistment age) was started, I was Wing Training Officer for Tasmania. I had no time or energy for research.

The war ended. I was able to get back to research, and Cramér's *Mathematical Methods of Statistics* was published, most opportunely for me. I was surprised and pleased when, soon after the war, early in 1947, I was invited to visit Columbia University and the University of North Carolina, and give courses in the Theory of Estimation and Testing Statistical Hypotheses, and in Non-parametric Methods. I was later invited to visit Princeton also.

In January 1948 I sailed from Sydney with my wife and three children in the *Marine Phoenix*, a United States troopship. We were setting off on a voyage of faith and hope. At that time, there was a dreadful housing shortage in the United States. Although we had been trying for nearly a year, and using every means we could think of, we had not succeeded in arranging for accommodation. It was a happy beginning when, on the first day out from Australia, we received a wireless message to say that a furnished house had been rented for us in New Canaan, Connecticut, about 50 miles from Columbia University, where I was to teach for a semester.

It was an adventure for me because I was completely inexperienced as a university visitor. Although I had taught mathematics for 26 years, I had never given a seminar, nor had I ever attended one. The staffs of Australian mathematics departments were too busy teaching to have time for seminars. There were too few mathematicians in any one department, there were no graduate schools, we had very few overseas visitors and we did not visit other universities in Australia. We did not have an Australian Mathematical Society until 1956. The Statistical Society of Australia was founded in 1962.

Finally, I was recovering from a severe illness. Five or six weeks before we were due to leave Australia, I was taken into hospital with a burst appendix. Fortunately, by that time penicillin had come into use and was available in Tasmania, and so I survived; but for more than a year afterwards I was subject to recurrent attacks of extreme fatigue, when I could no longer think. This never occurred when I was lecturing; but I fear some people may have found me rather dumb at times outside the class-room.

I had never discussed statistical theory with anyone who was not one of my students. M. G. Kendall's book [2] was the only place where I had seen printed references to my work.

We eventually landed at San Francisco after a voyage which was pleasant except that we were not allowed to go ashore at our first port, Auckland, because there was a polio epidemic there, and then we were not allowed ashore at any other port because the ship had called at Auckland. I soon started at Columbia, gave my two courses and went to the statistics seminars. I was impressed by the standard and the frequency of the seminars, and by the discussions, sometimes quite fierce, and also by the fact that the graduate students attended regularly. At times the seminars were beyond the comprehension of most of the students, but I thought that they learnt much by being constantly exposed to them.

I took with me to America, ready for publication, a paper in which I stated and proved the following theorem. *If* X, Y *are independent random variables, and* $aX + bY$, $a'X + b'Y$ *are also independent, where* a, b, a', b' *are all different from* 0, *then both* X *and* Y *have normal distributions*. The first seminar that I attended at Columbia was given by Loève, who had just arrived in the United States. At one point he stated a result which was a consequence of the theorem, and asked if anyone knew the theorem. I was the only person present who did. In the discussion that followed, someone said that he thought that something like it had been published by a Russian. Neither Loève nor I published it. It appeared a few years later, in a Russian journal, I think.

Abraham Wald was head of the statistics department. I had studied many of his papers, and learnt much from them. When I gave a seminar, he showed his mathematical power in the discussion afterwards. I realized that he had completely grasped what I had done. I admired him greatly, not only for his mathematical talent, but also for his great humility and kindness.

After the summer vacation, I moved to the University of North Carolina at Chapel Hill where I taught for two quarters. At both Columbia and Chapel Hill, after a course was finished, with the help of students, I produced a set of notes. The Chapel Hill notes on non-parametric inference became well known and much in demand. They were widely circulated, and were frequently referred to in the literature. Twelve years later, the Mathematical Centre in Amsterdam asked for permission to make copies for mathematicians and statisticians in Holland. Recently I was presented with a retyped copy by Nicholas Fisher, who had been visiting Chapel Hill.

It was not entirely my fault that the notes were not published in a more permanent and more accessible form. Shortly before I left the United States, I was visited by a representative of an American publisher, who asked if I had any manuscripts. I said yes, and told him about the non-parametric notes; but he was not interested. They would make only a small book, and his firm was interested only in big books. Could I expand the manuscript? No, not much. No deal. Twenty years later, a publisher wanted to print the notes; but I refused. I said they had done their work, and were now out of date.

The things in the notes that were most frequently referred to were the method of obtaining the asymptotic power of a consistent test, and the notion of asymptotic relative efficiency. Non-parametric tests had proliferated; but nothing seemed to be known about the powers of most of them.

Let θ be a real parameter of a probability distribution. Suppose that a test of the hypothesis $H_0: \theta=\theta_0$ against the hypothesis $H_1: \theta>\theta_0$, is reject H_0 if $T_n>K_n$, where T_n is a statistic, and n is the sample number. Suppose that the size of the test is exactly or approximately α, and in the latter case $\rightarrow \alpha$ as $n \rightarrow \infty$, so that

$$P_{\theta_0}(T_n>K_n)=\alpha_n \rightarrow \alpha \text{ as } n \rightarrow \infty.$$

The power function is

$$\beta(n,\theta)=P_\theta(T_n>K_n).$$

The test is said to be *consistent* if for every $\theta>\theta_0, \beta(n,\theta)\rightarrow 1$ as $n\rightarrow\infty$. Suppose that the test is consistent; let us investigate the behaviour of $\beta(n,\theta)$ for large values of n. Power functions are usually difficult to evaluate, and we mostly have to be content with approximations based on limit results. For fixed $\theta>\theta_0, \beta(n,\theta)\rightarrow 1$ as $n\rightarrow\infty$. Hence to get a limit value less than 1, we must consider a sequence (θ_n) of θ values such that $\theta_n>\theta_0$ and $\rightarrow\theta_0$ as $n\rightarrow\infty$. We try to determine this sequence so that as $n\rightarrow\infty, \beta(n,\theta_n)$ tends to a given limit between α and 1.

We assume that for some $h>0$ and for $\theta_0\leqslant\theta\leqslant\theta_0+h$, $a(\theta)$ and $w(n)$ exist, such that as $n\uparrow\infty$, $w(n)\downarrow 0$, and the θ_0 distribution of $[T_n-a(\theta_0)]/w(n)$, and the θ_n distribution of $[T_n-a(\theta_n)]/w(n)$ both tend to a distribution with a continuous distribution function F, as $n\rightarrow\infty$. We also assume that at θ_0, $a(\theta)$ has a θ derivative $a'(\theta_0)>0$.

Taking $\theta_n = \theta_0 + \lambda w(n)/a'(\theta_0)$, $\lambda > 0$, we can prove ([13], Chapter 7) that

$$\lim_{n\to\infty} \beta[n, \theta_0 + \lambda w(n)/a'(\theta_0)] = 1 - F(k-\lambda), \text{ where } 1 - F(k) = \alpha.$$

When n is not too small, this gives an approximation to the power function which is often sufficiently good for practical purposes. Ordinarily, we do not need to know the power function very precisely. Note that the sequence (θ_n) is introduced only for the purpose of obtaining a limit. It is not, as seems to be sometimes thought, a sequence of states of nature actually encountered.

Efficiency is a term imported from engineering and physics. It should be used only in a situation where there is an input and an output, and where at least one of these can be measured. Here we have a measurable input, the number of observations, and a measurable output, the power of a test against a specified alternative. If we have two tests of the same hypothesis at the same level α, and for the same power with respect to the same alternative, the first test requires a sample of n_1, and the second a sample of n_2, it is reasonable to define the relative efficiency of the second test with respect to the first as n_1/n_2. The limit of this when $n_2 \to \infty$ is the asymptotic relative efficiency, ARE. I was delighted when I discovered that the ARE of the Wilcoxon test compared with the t-test for samples from normal distributions was $3/\pi$.

5. Return to Australia; Sir Ronald Fisher

After two quarters at Chapel Hill, I spent a few weeks at Princeton, and then returned to Australia. My connection with agricultural field trials had ceased some years before the war. From time to time I was consulted about statistical problems by members of other university departments, and by people from the State Departments of Forestry and Health; but this work raised no problems of theory. It was in teaching that I met such problems. Two which I had with me for years and did eventually publish my solutions of, were about the relation between the behaviour of a distribution function in the neighbourhood of $\pm\infty$ and the behaviour of its characteristic function in the neighbourhood of 0, and about the Cramér–Rao inequality.

With regard to the first, the paucity of precise information irritated me. Let F be the distribution function, and ϕ the characteristic function. For $x \geqslant 0$, put

$$H(x) = 1 - F(x) + F(-x), \text{ the tail sum,}$$
$$K(x) = 1 - F(x) - F(-x), \text{ the tail difference.}$$

Let U and V be the real and imaginary parts of ϕ for real t,

$$\phi(t) = U(t) + iV(t).$$

We then find that U depends only on H, and V depends only on K. The functions H and K are only loosely connected. The only connections between them are (i) $H(x) \geqslant |K(x)|$ (ii) $H(x) \pm K(x)$ both non-increasing functions of x. A great simplification results if we consider U and V separately. This simplification enabled me to deal with the problem effectively, and I gave a statement of results at the fourth Berkeley Symposium in 1960 [10]. Proofs were published some years later [11].

I first met Sir Ronald Fisher when he made his first visit to Australia in 1953. He came to Tasmania for a few days over Easter, and stayed with us. We found him a charming guest. At a party you could lead him up to anyone, or anyone to him, make the necessary introductions and explanations, and leave him. He would quickly have the other person interested in talking with him.

When I asked him what he would like to do when we visited the university, he said that he liked talking individually with research students in a laboratory, and that he did not like to be asked to say a few words to a gathering. He surprised me by saying that he thought that the greatest effect of statistics on the world was in quality control, because we had so many complicated things like aircraft, with thousands of parts, that without quality control, their manufacture would be impossible.

I visited Stanford University for three quarters in 1957. While there, I was conned into writing a review article [9] on Fisher's *Statistical Methods and Scientific Inference*, which had just been published.

I did not try to write an ordinary review, which is intended to help a reader decide whether to buy the book, borrow it, or ignore it. I addressed myself to the people who had read the book, to those who would read it in the future, and to Fisher himself. He had spoken, and I wanted to continue the scientific dialogue. He didn't like my criticism of some of his arguments; but, as far as I know, he never attempted to reply, and he never raised the subject with me. Indeed, for some years he could hardly bear to speak to me when we met from time to time in various places. However, we did eventually become friendly again without ever discussing the matter.

The beginning of fiducial probability, fiducial distributions, and confidence intervals was Fisher's 1930 paper, "Inverse Probability" [1]. The basic idea in his argument is extremely simple; but no one seems to have thought of it before Fisher. We may put it as follows. Let T be an estimator of a parameter θ. Given a number k between 0 and 1, define $K(\theta)$ by

$$P\{T \geqslant K(\theta)\} = k. \tag{1}$$

If $K(\theta)$ is an increasing function of θ, the inequality $T \geqslant K(\theta)$ is equivalent to $\theta \leqslant K^{-1}(T)$, where K^{-1} is the inverse function of K. Thus (1) can be written

$$P\{\theta \leqslant K^{-1}(T)\} = k. \tag{2}$$

It is hardly possible for present-day students of statistics to appreciate the liberating effect of the mathematically trivial step from (1) to (2), a simple twisting of an inequality. Neyman expressed himself as follows [3].

> The possibility of solving the problems of statistical estimation independently from any knowledge of the *a priori* probability laws, discovered by R. A. Fisher, makes it superfluous to make any appeals to the Bayes theorem.... The present solution means, I think, a revolution in *the theory* of statistics.

I was enthusiastic. I remember the first time I talked about it in my lectures. The next time I met that class, I walked over to a student and asked where I had got to in the last lecture. He showed me his notebook. I was interested to see that across the page he had printed in very large capitals BETTER THAN BAYES. He had evidently got the message.

The interval $(-\infty, K^{-1}(T)]$ is a confidence interval, at confidence level k; but Fisher did not use this term. He called $K^{-1}(T)$ an upper fiducial limit. From this beginning he went on to the fiducial distribution of an unknown parameter. Neyman, starting from here, developed the theory of confidence intervals.

6. The Cramér–Rao Inequality

The second problem that bothered me for years was about the Cramér–Rao inequality

$$\frac{h'(\theta)^2}{V(S)} \leqslant I = \int \frac{f'^2}{f} d\mu.$$

Here f is a probability density that depends on θ, and the prime denotes differentiation with respect to θ. V is the variance of S. It is assumed that

$$\int f' \, d\mu = \frac{d}{d\theta} \int f \, d\mu = \frac{d}{d\theta}(1) = 0.$$

S is a statistic with a finite mean value $h(\theta)$ for θ in an open interval N. We say that S is regular if h has a θ derivative given by

$$h'(\theta) = \int S f' \, d\mu.$$

The inequality is true if S is regular.

Which statistics are regular? We mostly want to apply the Cramér–Rao inequality to statistics that we do not know, and so the regularity conditions should ask as little as possible of the statistic S, which we may not know, and should be mainly concerned with the family of probability measures P_θ, which we know completely.

Fisher seems to have been the first to encounter $I = \int f'^2/f \, d\mu$, He called it the intrinsic accuracy of the distribution, and later, the amount of information in an observation. The former name has dropped out of use. I is now usually called the information, or the Fisher information; but this is not a good name. Suppose that in some neighbourhood of $\theta = 0$, $I = \int (f'^2/f) \, d\mu$ is finite and greater than 0. Put $\phi = \theta^3$. The density function is

$f(.,\phi^{1/3})$, and $df/d\phi = f'\,d\theta/d\phi = \phi^{-2/3}f'/3$. Hence

$$\int (df/d\phi)^2/f\,d\mu = \phi^{-4/3}I/9.$$

This is ∞ at $\theta = 0$. Similarly, if $\psi = \theta^{1/3}$,

$$\int (df/d\psi)^2/f\,d\mu = 9\psi^4 I,$$

which is 0 at $\theta = 0$. Thus, at $\theta = 0$, we have finite non-zero "information" about θ, infinite "information" about θ^3, and zero "information" about $\theta^{1/3}$.

$$I_0 = \int \frac{f_0'^2}{f_0}\,d\mu = \int 4 \lim_{\theta\to\theta_0} \left(\frac{f^{1/2} - f_0^{1/2}}{\theta - \theta_0} \right)^2 d\mu.$$

If this is finite and equal to

$$4 \lim_{\theta\to\theta_0} \frac{\int (f^{1/2} - f_0^{1/2})^2\,d\mu}{(\theta - \theta_0)^2}$$

we say that the P_θ family of probability measures is *smooth* at θ_0, and I_0 is the *sensitivity* of the family at θ_0. It turns out that the Cramér–Rao inequality is of interest only when the family is smooth. When this is so, every statistic with a variance which is bounded for θ in some neighbourhood of θ_0, is regular at θ_0. That would seem to be the last word on this famous inequality.

7. Concluding Remarks

In conclusion, I wish to say that I take the view that the aim of the theory of statistical inference is to provide a set of principles which help the statistician *to assess the strength of the evidence* supplied by a trial or experiment, for or against a hypothesis, or *to assess the reliability of an estimate* derived from the result of such a trial or experiment. In making such an assessment we may look at the results to be assessed from various points of view, and express ourselves in various ways. For example, we may think and speak in terms of repeated trials, as for confidence limits, or for significance tests, or we may consider the effect of various loss functions. Standard errors do give us some comprehension of reliability; but we may sometimes prefer to think in terms of prior and posterior distributions, or to consider a fiducial distribution. All of these may be helpful, and none should be interdicted. The theory of inference is persuasive rather than coercive. It does not provide a set of infallible rules which are to be applied automatically.

If the theory of inference is to be of use to the practising statistician, it must be clear and simple, and much statistical theory is not. Regulatory conditions for theoretical results are often flung down with scant regard for the possible user. They are often too strong, and they are often difficult to verify in actual cases. In the development of a theory, simplicity usually comes last; but we must continually strive for it.

Finally I want to stress the necessity to be honest. We must be scrupulous in the use of terminology so as not to mislead by suggesting more than can be justified. We must avoid the indiscriminate use of terms like *efficiency*, *information*, *best*.

Publications and References

[1] Fisher, R. A. (1930) Inverse probability. *Proc. Camb. Phil. Soc.* **26**, 528–535.

[2] Kendall, M. G. (1946) *The Advanced Theory of Statistics*. Vol. II, 1st edn. Griffin, London.

[3] Neyman, J. (1934) On the two different aspects of the representative method. *J. R. Statist. Soc.* **47**, 558–625.

[4] Pitman, E. J. G. (1936) Sufficient statistics and intrinsic accuracy. *Proc. Camb. Phil. Soc.* **32**, 567–579.

[5] Pitman, E. J. G. (1937) Significance tests which may be applied to samples from any populations. *J. R. Statist. Soc. Suppl.* **4**, 119–130.

[6] Pitman, E. J. G. (1937) Significance tests II. The correlation coefficient test. *J. R. Statist. Soc. Suppl.* **4**, 225–232.

[7] Pitman, E. J. G. (1938) Significance tests III. The analysis of variance test. *Biometrika* **29**, 322–335.

[8] Pitman, E. J. G. (1939) The estimation of the location and scale parameters of a continuous population of any given form. *Biometrika* **30**, 391–421.

[9] Pitman, E. J. G. (1957) Statistics and science. *J. Amer. Statist. Assoc.* **52**, 322–330.

[10] Pitman, E. J. G. (1960) Some theorems on characteristic functions of probability distributions. *Proc. Berkeley Symp. Math. Statist. Prob.* **2**, 393–402.

[11] Pitman, E. J. G. (1968) On the behaviour of the characteristic function of a probability distribution in the neighbourhood of the origin. *J. Austral. Math. Soc.* **8**, 423–443.

[12] Pitman, E. J. G. (1978) The Cramér–Rao inequality. *Austral. J. Statist.* **20**, 60–74.

[13] Pitman, E. J. G. (1979) *Some Basic Theory for Statistical Inference*. Chapman and Hall, London.

[14] Scott, R. A. (1932) Some methods of testing potato yields. *Tasmanian J. Agric.* **3** *Suppl.*, 2–15.

[15] Yates, F. (1935) Complex experiments. *J. R. Statist. Soc. Suppl.* **2**, 181–247.

4
STATISTICIANS IN DESIGN AND COMPUTING

R. L. Anderson

Richard L. Anderson was born at North Liberty, Indiana on 20 April 1915. After completing his high school education there, he obtained his AB degree at DePauw University in 1936, and his Ph.D. in Mathematics, Statistics and Economics at Iowa State College in 1941.

He began his academic career at North Carolina State College (later University), and rose from instructor to full professor in his 25 years of service there to 1966. He was away from Raleigh in 1944–45 when he consulted with the Army and Navy at Princeton, in 1950–51 when he visited Purdue, and during 1958 when he spent some time at the London School of Economics. In 1967, after a year as visiting research professor at the University of Georgia, he took up the chairmanship of the newly established Department of Statistics at the University of Kentucky, Lexington. He withdrew from this position in 1979 and is now serving as Assistant for Statistical Services to the Dean, College of Agriculture, University of Kentucky.

Professor Anderson has long-standing interests in experimental design, regression methods, variance components and time series analysis and their application to agricultural, industrial and operational problems. He has taken an active role in statistics both in the USA and internationally. He is a Fellow of the American Statistical Association (ASA), of the Institute of Mathematical Statistics and of the American Association for the Advancement of Science. He has been President of the Eastern North American Region of the Biometric Society, is a member of the International Statistical Institute, and is currently President-Elect of the ASA. He is widely consulted, and has been invited to advise on statistical matters in Japan, Sweden and India.

Professor Anderson is the joint author of the well-known book *Statistical Theory in Research* and of numerous papers in statistics. He is married, and has one son and one daughter. His recreations have been softball, tennis, bowling and his family.

My Experience as a Statistician: From the Farm to the University

R. L. Anderson

1. The Beginning in Indiana

I was born in 1915, and reared on a general purpose farm near North Liberty, Indiana; I lived through the farm depression of the 1920s which became the general depression of the 1930s. I can remember counting the clover seed bags in 1932 to see if there would be enough cash for me to go to DePauw University, where I had a tuition scholarship; I can also recall the letters from my mother, stating that here was half the money for my room and board that month and the rest would come later. Although I did wait on tables to supplement my income, my parents made every effort to provide enough money for me to devote most of my time to studies and not to outside work. Interest in statistics came from two sources: observing the natural variability in crop and livestock production on the farm and then electing a course in mathematical statistics under Professor Greenleaf in my senior year at DePauw.

I can remember the efforts my father and I made to improve crop yields, including our successful use of green manure whereby we plowed under sweet clover to increase wheat yields; my father took great pleasure in explaining how we were able to obtain 35 bushels to the acre that summer. I can still remember clearly the people stopping by to see the abundance of wheat shocks in that field (this was before combines were used for grain harvesting). But anyone who longs for a return to "the good old days" probably does not know what the good old days entailed, especially in the snow belt which spread from Lake Michigan around to South Bend. Those were the days of hand and foot labor from 5 a.m. to 8 p.m., when going to school was the vacation. There was never any question of my participating in any activity at college that might have forced my dismissal back to the farm.

My high school activities were quite limited. I had the misfortune of skipping two elementary grades (a common practice in those days) so that I was at least two years younger than the other members of my high school class. My small stature prevented my playing basketball, a major handicap in Indiana, where basketball dominated local conversation throughout the winter. When I was awarded a DePauw scholarship, I discovered that I could not meet the entrance requirements. In my sophomore year, when I tried to enroll in plane geometry, I was urged to take the agriculture curriculum because I was a farm boy. When my father learned that this advice might prevent my attending a non-technical university, he almost had the agriculture program eliminated from the local high school curriculum. To make a long story short, I took a postgraduate year of high school, during which I studied two years of Latin (one in summer), plane and solid geometry and advanced algebra; fortunately DePauw granted me a scholarship the next year.

At DePauw I majored in mathematics with a strong minor in history; my interest in the latter persists to this day. My staunch Republican father never forgave DePauw for converting me to the Democratic Party. My most persuasive history teacher, Professor Crandall, was a solid Democrat, who took great delight in pointing out how the Republicans maintained their power after the Civil War by "waving the bloody flag of the rebellion." He delighted in the success of Franklin D. Roosevelt. I was also introduced to economics at DePauw. At the end of my studies I received an Indiana high school teacher's license; fortunately I had not located a position when an unexpected chance to pursue graduate work came my way.

2. Graduate Work at Iowa State

During the summer of 1936, one of the mathematics professors at Iowa State College died, and the department decided to replace the position by a number of teaching assistantships. Since the heads of the mathematics departments at Iowa State and DePauw had been fellow graduate students, the former contacted DePauw for possible candidates; I was one of those informed. Believe it or not, this was the first time I had learned about graduate school and teaching assistantships; what a change to today, when graduate departments conduct recruiting drives for good graduate students. One of the attractive features at Iowa State was the chance to pursue statistics and economics in addition to mathematics.

The transition from high school to college had not been difficult for me, primarily because of excellent high school training; however, the transition to graduate school was more difficult, primarily because of a course in complex variable function theory and Professor Snedecor's course in statistical methods. The transition from mathematical statistics to applied statistics was also not an easy one. George Snedecor regarded me as a diehard mathematical statistician for many years. I recall that one of my major problems was covariance, and I never understood the basis for it until I took a course involving least squares regression from Professor W. G. Cochran several years later.

I minored in economics and was fortunate to become a student of Gerhard Tintner, an outstanding econometrician. Since his major research area was time series, where he popularized the variate difference method, which has been widely used in detrending, I naturally fell into this area. My master's thesis, presented in 1938, was concerned with the analysis of a number of agriculture price series (seasonal, trend and cyclical components) and was entitled "Hidden Periodicities in the Price Fluctuations of Selected Agricultural Commodities." I discovered that there did not exist a test of the existence of a real period for correlated observations; R. A. Fisher had obtained a test for harmonic analysis, but it was based on independent observations. This led me to develop a test for independence, which culminated in my doctoral dissertation, "Serial Correlation in the Analysis of Time Series" (1941). This research was improved tremendously by a suggestion from Cochran that I use his theorem on quadratic forms. Whereas the dissertation was concerned with data with known mean, my *Annals* article [1] emphasized subsequent results for deviations from the sample mean. The serial correlation (now called autocorrelation) coefficient for lag 1 and n observations was

$$r_1 = \frac{\sum_{i=1}^{n} X_i X_{i+1} - \left(\sum_{i=1}^{n} X_i \right)^2 \Big/ n}{\sum_{i=1}^{n} X_i^2 - \left(\sum_{i=1}^{n} X_i \right)^2 \Big/ n}$$

where X_{n+1} was equal to X_1, following the circular definition suggested by Harold Hotelling.

After my first year at Iowa State, my assistantship was transferred to the Statistical Laboratory where I had charge of the laboratory section of the statistical methods courses; in the latter role, I had to learn to use the mechanical Hollerith computer. I was the second Ph.D. graduate in statistics at Iowa State, Holly Fryer having preceded me by a year. Both of us actually obtained degrees in mathematics, since the Ph.D. in Statistics had not been approved when we started our programs. Most of my formal training in statistics came after Cochran joined the Iowa State faculty in 1939; my rewarding experiences with him are detailed in the obituary [12] I prepared for *Biometrics*. Suffice to say that my research and consulting in linear models, including variance components, and experimental and survey design were based on the concepts presented by Professor Cochran.

My teacher in experimental design procedures was Gertrude Cox, who was also a fellow graduate student. Unfortunately, Gertrude was never given time off from her consulting and teaching duties to write a dissertation. Then in October 1940 she surprised everyone by accepting the newly created position of Head of Department of Experimental Statistics at North Carolina State College in Raleigh. After enduring many years of Iowa's cold winters, I told Gertrude that if she had a vacancy for a mathematical statistician, I was interested in coming south. This desire was strengthened when my softball catcher, Bob Monroe, joined Gertrude early in 1941 to take charge of her computing facility (similar to that at Ames). From this you will infer that I had become a softball pitcher at Ames, an avocation which continued for many years.

3. Going South to North Carolina State; Army–Navy Research at Princeton

Before Gertrude could find a position for me at Raleigh, I received an offer of an instructorship in mathematics at North Carolina State, which I accepted with the understanding that I could transfer to statistics if a position became available there. This transfer occurred in 1942, when I became an Assistant Professor; however, I continued to teach mathematics in the Army Specialized Training Program (ASTP) which was conducted at North Carolina State for talented members of the U.S. Army.

The entrance of the United States into World War II in December 1941 was a traumatic event for many of us. I was deferred from military service to handle the ASTP teaching but took over the computing facility when Bob Monroe entered the army. In 1944, Cochran and Sam Wilks persuaded me to join the latter's Statistical Research Group at Princeton, where I remained until October 1945. My first project concerned research on a naval air project, for which I analyzed many sets of flight data. Unfortunately these data contradicted some results obtained in laboratory tests at MIT. We discovered that the naval officers who had conducted the flights had failed to activate the testing device; hence, all of the results had been random. We were in a quandary as to how to write our last report, which would contradict everything in the other reports, without condemning the naval personnel, who had to approve the report. It was then that I came to respect Sam Wilks for his ability to condemn someone without actually doing so; I regret that I do not have a copy of our letter (which had a Secret classification at the time).

I then turned to a project with Alex Mood involving the Army Corps of Engineers, which was conducting experiments on the effectiveness of explosive devices in clearing land mines. This was my first experience of the use of simulated experiments. With the aid of technicians we conducted many simulated bombing runs, which demonstrated the fact that this technique could not be used successfully. Then we helped set up actual experiments using liquid explosives at the A. P. Hill Military Reservation. One interesting experiment may be worth describing. The plan was to lay a plastic hose filled with explosive on the ground and insert indicator mines in the ground on each side of the hose at specified distances from it. The explosive in the hose would be detonated and the shock measured on the indicator mines; presumably shocks above a certain level would have detonated actual mines. The purpose of this experiment was to detect the effective range of the explosive device. The officer in charge did not wish to use simulator mines adjacent to the hose because he feared they would be destroyed by the blast. After all, everyone knew the shock could be heavy adjacent to the hose. However, Alex and I were insistent that mines should be placed adjacent to the hose. The mines were placed that evening and the blast was made the next morning; fortunately the ground froze sufficiently hard that night for the blast not to blow the adjacent mines from the ground. The astounding result was that they received only a very small shock; apparently the shock wave rises from the blast and comes down several feet away from the source. This indicated that the only effect of the blast would be felt in the cleared-out trench and not the area outside the trench; sufficient to say, the project officer never questioned us again regarding the proper design of an experiment, and we were consulted prior to every experiment thereafter. I should add that the physicists soon discovered that there was a theoretical reason for these results.

I have fond memories of those 18 months at Princeton. Wilks had gathered a collection of excellent statisticians, many being his own graduate students. Ted Anderson and I worked on serial correlation, which resulted in a joint article [14]. Will Dixon and I shared an office and subsequently collaborated on his BMD-computer programs. But the friendships with these and other members of the group have endured. It was during my stay at Princeton that Gertrude Cox met me in New York to outline the plans for the Institute of Statistics, with a graduate program at Raleigh under Cochran and one at Chapel Hill under Hotelling. With these bright prospects ahead, I eagerly returned to Raleigh in October 1945.

4. Return to Raleigh

The most important event in my life took place in the winter of 1945, when Mary Turner (a truly gracious and charming southern belle) agreed to share the rest of her life with me. Gertrude Cox, who became the godmother of our two children, repeatedly reminded me that my demeanor improved tremendously after our marriage in January 1946.

It was an exciting statistical era at North Carolina State. Starting in 1944, Gertrude, with financial assistance from the General Education Board, embarked on her adventure to promote the development of statistics in the southern part of the United States. A series of work conferences were held: Plant Science, Quality Control, Agricultural Economics, Animal Sciences, Plant Genetics and Taste Testing. For me, these provided an excellent opportunity to become familiar with research efforts in a variety of fields; this information was especially valuable in transforming me from a theoretical to an applied statistician. In addition to these conferences, Gertrude also secured funds to sponsor two summer conferences in the mountains of North Carolina (Junaluska in 1946 and Blue Ridge in 1952); these were attended by prominent statisticians from many institutions and countries, and were instrumental in establishing the Institute of Statistics as an international center. I recall the Summer School at Raleigh in 1946, which preceded the Junaluska Conference and featured R. A. Fisher. I was persuaded to hold a rump session of students of Fisher's class, in which I tried to explain in more detail what he had been expounding in his lectures. I think I benefited more from these sessions than did the students, because I was forced to try to understand what the great man was trying to get across.

Mary got her first extensive contact with that strange breed called statisticians that summer. One of the most memorable events was escorting Professor Fisher to historic eastern North Carolina and its sandy beaches. Despite warnings from Mary, Fisher insisted on riding the waves with his glasses on; after losing his glasses in the surf and getting one of the worst sunburns in history, he was returned to a very irate Gertrude, who blamed everything on us. By the way, did anyone ever succeed in telling Fisher what to do?

Apart from statistics, this was the time that big-time basketball (Hoosier variety) was introduced to North Carolina. Everett Case came from Indiana to start that frantic scramble for basketball stardom with which I had been so familiar back home. The statistics faculty and graduate students were sports enthusiasts, from being basketball and football fans to participating actively in tennis, bowling, volleyball, softball and golf. I remember the night that Ralph Comstock did the best bowling of his life when we won the faculty championship. And who will ever forget the famous Patterson Hall spring golf tourney; all sorts of special prizes were concocted so that duffers like me would have a chance to win something. Softball had become almost a disease with me as a result of our success at Ames and volleyball and tennis were equally exciting.

It was during this period that I also became actively involved with sample surveys, when I served as the statistical advisor for a number of regional home economics and agricultural economics projects. In connection with the latter, I should mention that most of the original members of Gertrude's Raleigh faculty were assigned to consult with specific departments in the College of Agriculture (we were a department in that College until 1960); I was assigned to Agricultural Economics, and became a member of almost all their Ph.D. committees which involved statistical analysis. After Cochran left North Carolina State in 1949, I had to take over a number of his students and assumed the role of Graduate Administrator soon thereafter. My last serious involvement with time series research was with Geof Watson's research, most of which was carried on at the University of

Cambridge, where he also worked with Jim Durbin. This research produced the popular Durbin–Watson test for serial correlation of residuals from a fitted least squares regression, in addition to Geof's other work. There were three Ph.D. dissertations (Al Finkner, Emil Jebe and A. R. Sen) plus a number of MS theses in the sample survey area. I continued to work with Al on sample surveys until Dan Horvitz joined the North Carolina State statistics faculty.

Our first child, Kathy, was born in August 1950. That fall and winter the three of us spent a semester at West Lafayette, Indiana, where I worked at Purdue's Statistical Laboratory with Carl Kossack. It was an excellent opportunity for Mary to become acquainted with the Anderson clan, since North Liberty was only 100 miles away.

After we returned to Raleigh, I completed my part of the Anderson–Bancroft book, *Statistical Theory in Research* [15]. Although Cochran had left for Johns Hopkins, he had been quite helpful in the early stages, when we were developing the lecture notes for a theory course for the statistics graduate students. These notes were combined with a similar set developed by Ted Bancroft at Iowa State to produce the book. I have been chastised by my family, my colleagues and my former students for failing to rewrite my part of the book; I demurred because I did not want to go through another period of working nights and weekends on the book instead of spending my time with my family. This became even more important when our last child, Bill, born in May 1953, became old enough to play Little League baseball, and both Kathy and Bill became public school students. Nevertheless I am pleased that the book even now, after almost 30 years, seems to be useful to so many people.

5. Spreading the Gospel

One of Gertrude Cox's main objectives for the Institute of Statistics was to develop strong statistical programs throughout the South; this objective was sometimes referred to as "spreading the gospel according to St. Gertrude." She persuaded the Southern Regional Education Board (SREB) to establish a committee on statistics. I attended the developmental meetings which led to this decision. In order to find out what we had to start with, funds were obtained for Gertrude, Jack Rigney and me to contact southern universities which had some form of graduate training in statistics. In 1952 Mary, Kathy and I traveled throughout Georgia, Alabama, Mississippi, Tennessee and Kentucky. Based on this inventory, the SREB Committee decided there were three basic needs in statistics for the South: (1) to develop a series of summer sessions to upgrade the basic level of statistical competence; (2) to provide consulting help on an interim basis; (3) to encourage the development of more statistics departments. Some of the latter should develop strong graduate programs.

The SREB Committee requested Virginia Polytechnic Institute (VPI), North Carolina State, Oklahoma A&M and the University of Florida to conduct a series of six-week regional summer sessions on a rotating basis. We had offered a special summer session in 1951 and VPI in 1951 and 1952; the regional program started at VPI in 1954 and continued through 1972, with VPI and six southern universities not in the original group conducting the sessions from 1966 through 1972. I had the pleasure of teaching my Advanced Experimental Statistics course at five summer sessions. This course covered topics introduced by Cochran when he was in Raleigh, and gradually introduced more and more topics in the design and analysis of variance component and mixed models. These summer sessions resulted in a continuous set of courses in basic statistical theory and analysis which could form the basis for an MS program in statistics at any of the co-operating institutions. The courses were supplemented by selections from advanced courses in all areas of statistics and probability. These summer sessions contributed a tremendous amount to the upgrading of statistical competence throughout the southern USA and even beyond.

By 1967, it became obvious that most of the universities had sufficient summer programs of their own so that the regional summer session no longer had a high priority. Hence in 1968, the SREB initiated a series of one-week Summer Research Conferences, the first at Montreat, North Carolina. These conferences have become a regular feature in the activities of the statistical community. It has been a pleasure to be associated with the SREB Committee on Statistics since its inception; the Committee has justified every hope of Gertrude Cox and Boyd Harshbarger, who were the prime movers in getting it started. Among its numerous benefits is the fact that the camaraderie among southern statisticians in the USA is unique.

The statistics group at North Carolina State conducted a series of short courses for government and industrial employees. In 1956, we held a one-week Conference on Experimental Designs in Industry; this featured many outside speakers as well as some of us from the Institute. A highlight of this conference was the election-night parties; Mary and I entertained the sad Democrats while Gertrude took care of the jubilant Republicans. I still recall Besse Day sitting on the floor before our television set in complete disbelief at the injustice of it all!

I should mention the trips that several of us from Raleigh made to experimental stations in the South to help them become acquainted with modern procedures in conducting sample surveys, designing experiments, and collecting and analyzing data. I remember with pleasure trips to Mississippi State, Auburn, Clemson and Florida.

Another method of spreading the gospel was to get us involved in national and international projects. I became involved in the development of high-speed computers, even though I have never learned to program one of the beasts. I remember a session held at Endicott House, near Boston, in 1956. As chairman of an *ad hoc* High Speed Computer Committee of the

Institute of Mathematical Statistics, following the questionnaire distributed in 1956, I became aware of the low regard in which computers were held by IMS members. It became apparent that most of them regarded the computer as simply a device to speed up existing operations, rather than one which would enable them to do things which could not be done at all without it. Fortunately, most statisticians now perceive the computer as an essential element in their research, consulting and teaching programs, but I believe it took them too long to arrive at that conclusion.

In 1958, I took my first international trip, when the Anderson family spent six months in London. I had two main objectives: (1) to learn more about the work of Professor Phillips at the London School of Economics in connection with a Dynamic Economics Processes project, sponsored by the Ford Foundation (and jointly researched at the LSE and the Institute of Statistics), and (2) to work with W. R. Buckland on the development of the *International Journal of Abstracts in Statistical Theory and Method* for the International Statistical Institute (ISI).

The latter effort was quite successful, as demonstrated by the fact that the abstracting journal has continued to be published by the ISI. We set up an article index and a procedure for abstracting most articles which developed new or improved theory or methods. After returning to Raleigh, I devoted many hours to the problem of securing abstracters in the United States.

I am afraid that I never fully understood Bill Phillips' electrical engineering formulation of economic policy. We sent two more statisticians from Raleigh to England to work on this aspect of the project, but the joint effort was not very productive. I had several discussions with Bill on the relationship between unemployment and economic progress. (Most of you are probably familiar with the Phillips curve.) I argued that length of unemployment was probably more important than a simple count of those unemployed; this becomes even more important today, when unemployment benefits are provided for only a certain period of time.

There were many side benefits from this 1958 project. Our daughter benefited greatly from a learning experience at the American School in London, and we all learned something about England and Western Europe as a result of two trips to the Continent, including one to Scandinavia and one to the World's Fair and ISI meeting in Brussels, followed by a trip to Switzerland with Gertrude. I had an opportunity to discuss research ideas with many outstanding statisticians; this included presenting a paper on variance components at the ISI meetings and discussing simultaneous economic relationships with Herman Wold at his summer retreat in southern Sweden.

In 1965, I was invited to participate in an NSF-sponsored trip to attend a meeting in Tokyo on sampling of bulk materials. I persuaded Mary to hire someone to stay with Kathy and Bill, so that she could go with me. It was truly one of the highlights of our travels. We met so many charming and helpful people, who made sure that we became as familiar with Japan as one can in such a short period of time.

6. Developing a Graduate Program

I became Graduate Administrator for the Statistics Department at North Carolina State in 1953. At this time there were approximately 30 graduate students in the program at Raleigh. This number was to increase each year until it reached over 100 in 1966. I became a member of the Administrative Board of the Graduate School in 1960. There was continuing difficulty in allocating prospective Ph.D. candidates to dissertation directors in a department which was dominated by a faculty hired primarily as consultants. Despite the many headaches of the position, it carried many benefits, the most important being the opportunity to associate with and counsel so many excellent graduate students. I especially remember the picnics held at our house and back yard for graduate students who had completed their degrees. Those were the golden days of graduate education. The remainder of this section will be devoted to the research programs with which I was involved as a part of the graduate program.

6.1. Variance Components

While at Purdue in 1950, I was asked to recommend a design to compare the sources of variation in the production of streptomycin. There are five stages in the production and assay process: the initial incubation stage in a test tube (or slant, as it is generally called), a primary incubation stage in a petri dish, a secondary incubation stage, a fermentation stage in a bath, and the final assay of the amount of streptomycin produced; I called this a five-stage nested design. I demonstrated that even with as many as 80 assays in a balanced sampling plan, we could use only five test tubes (with two samples in each stage thereafter) so that the standard errors of the estimated slant and primary inoculation variance components would be almost as large as the components themselves. I recommended the use of a systematic non-balanced sampling plan, which I called a "staggered design."

When we returned to Raleigh, I decided to pursue some research on optimal designs to estimate variance components. Phelps Crump, whose brother Lee had started his excellent fundamental research in variance component estimation at Iowa State, agreed to work on the problem for a two-stage nested design. His monumental work, which unfortunately was not published until 1967 [17], was the first of a series of 10 Ph.D. dissertations, the last of which was completed in 1980: Gaylor and Bush [20] at North Carolina State and Muse [24], Thitakomal and Stroup [26] at Kentucky on two-way classification designs; Prairie at North Carolina State and Schwartz at Kentucky on three-stage nested designs; Kussmaul at North Carolina State on composite designs [22], Sahai at Kentucky on both nested and classification designs with both regular and Bayesian estimation [25]. One of the important recent developments has been the improved efficiency of maximum likelihood over standard analysis of variance estimators for non-balanced designs. Computing procedures have been developed for the BMDP programs and some SAS programs are available. I have developed an iterated weighted least squares procedure which is quite useful for some of the non-balanced designs. I am finally writing a book on variance components. Many of my ideas on this topic were developed in discussions with Cochran while he was in Raleigh and thereafter. I published one article [2] on their use for time series data in 1947; presentations have been made at a number of conferences: 1954, Quality Control [3]; 1958, ISI [6]; 1961, Army Research and Development [7]; 1965, Bulk Sampling in Japan [8]; 1967, International Biometric Conference (IBC) [9]; 1973, Statistical Design and Linear Models Symposium [11], and at many university seminars.

6.2. Linear-Plateau Models

The Institute of Statistics had a long-term co-operative research contract with the Tennessee Valley Authority (TVA) in conjunction with research workers in soils to determine optimal fertilizer rates for corn. The proceedings of two symposia sponsored by TVA were published by the Iowa State College Press in 1956 and 1957; I have a chapter in each on this topic [4], [5]. Data collected in 1955, 1956 and 1957 formed the basis for a number of research reports and dissertations, but I was never satisfied with the estimated optimal levels of applied nutrients because of the observed plateauing of yields, often at moderate fertilizer levels. As a result, the usual quadratic production models tended to produce overestimates of the optimal fertilizer levels. In the meantime Larry Nelson joined the Statistics Department at North Carolina State to work with the research workers in soils; he and I revived the analysis of the TVA data. In a moment of weakness in 1969, we promised Gertrude Cox to prepare a paper on this topic for a session at the 1971 ISI meeting.

This paper [18] initiated a joint research program which is still in progress. We have developed a series of linear-plateau models, which seem to approximate more adequately the response patterns of crops to fertilizer. These models were first presented at the 1974 IBC in Romania and published in *Biometrics* [19] in 1975; some suggested improvements were presented at the 1979 IBC in Brazil. We have received an unprecedented number of requests for reprints of these articles, from many countries including some in Eastern Europe, but all from agricultural and biological research workers. A notable adaption of these procedures is now being made in drug research, where the plateau is at low doses, i.e., a threshold. A recent Ph.D. dissertation (Neidert) has initiated a comparison of various estimation procedures by use of simulated experiments for known populations. Much more work is needed in this area.

6.3. Use of Prior Information in Regression

During the period 1962–67, I consulted with engineers and statisticians at the Air Force Missile Test Center, Patrick Air Force Base, Florida. One of their most intriguing problems has been that of determining the Best Estimate of Trajectory (BET) based on pooling information obtained from a number of range instrumentation systems, simultaneously tracking a given missile. Each system has a number of systematic calibration biases which change from flight to flight, but which may be more or less stable for a given flight. Since the systems usually have entirely different electronic components, different systematic biases must be determined for each system. One approach to the removal of the systematic biases (designated as low-frequency errors by the electrical engineers) is the use of some hardware-oriented calibration schemes. Another approach, and the one to be discussed here, is to estimate the magnitude of these bias parameters simultaneously with trajectory coordinates and velocities by standard weighted regression procedures; this is often called the "self-calibration" solution.

In this weighted regression process, the engineers assume that certain of the bias parameters have a prior distribution with means and variances based on static tests and previous operational flights. One of my graduate students, E. L. Battiste, studied the effect of using biased priors in estimating the parameters of linear models with two highly correlated independent variables and I extended the results to three independent variables [16]. Later W. C. Gregory considered the use of priors for non-linear models, which are more typical of missile data [21]. At the same time, Hoerl and Kennard were developing their "ridge regression" procedures, which are also based on the use of prior information.

An overall evaluation of these procedures is fraught with danger, but the following rationale may be applicable:

(1) Suppose that the experimenter obtains an estimate of a parameter on a given day or at a given place, but its indicated variance is very large, either because of an ill-conditioned matrix, poor measures or poor experimental conditions.

(i) A large variance indicates that the estimator ($\hat{B}$) may have a large deviation from B.

(ii) If there is prior information on B, say B_0, and this poorly estimated $\hat{B}$ deviates far from B_0, the experimenter would like an estimating procedure which would give a pooled estimator close to B_0.

(2) On the other hand if the indicated variance of $\hat{B}$ is small, the pooled estimator should be close to $\hat{B}$.

(3) If $\hat{B}$ and B_0 are nearly equal, all linear pooling procedures will give nearly identical results.

(4) We know that the prior, B_0, will be a biased estimator of B. Hence it will be desirable to use as a weight for B_0 a quantity somewhat less than the reciprocal of its prior variance, in order to minimize the average mean square error. We note that if there is little bias in B_0, a slightly larger prior variance will not materially reduce the gain due to the use of the prior; however, if there is a material bias, a value of prior variance which is too small can be disastrous.

(5) Finally in missile experiments, as in many others, the non-linearity of the model plus the large number of parameters to be estimated necessitates the use of a standard computer program which will produce reasonably good results under a variety of conditions.

6.4. Other Research Topics

My colleagues, students and I have merely scratched the surface of a number of research topics, which merit much additional investigation both in the development of better theory and the use of additional simulation projects, such as:

(a) The effect of misclassification in contingency tables and other forms of categorical data [23].

(b) The optimal designs and estimators for parameters in non-linear models.

(c) The optimal procedures in selecting predictors in multiple regression [13].

7. Other Activities at North Carolina State

North Carolina State was a leader in the development of the use of high-speed computers in universities. We entered the modern age in 1956, when the IBM 650 was installed. The important SAS programs started in the Statistics Department with encouragement from the Southern Agriculture Experiment Stations. Then came the IBM 360 systems, with concentration at the Research Triangle Institute (RTI). The latter had been developed with the aid of a number of us working with Gertrude Cox, who resigned from the university to become Director of the Statistics Section of RTI in 1960.

A strong graduate program in biomathematics under the leadership of H. L. Lucas was developed. I remember the impetus provided by an outstanding conference held in western North Carolina in 1961.

I became the Chairman of an Operations Research Advisory Committee, which developed a combined program with departments on both the Raleigh and Chapel Hill campuses; a Master's program was approved in 1964 and directorship of the North Carolina State part of the program was placed in the Department of Industrial Engineering.

I recall the many hours spent developing plans for a School of Statistics at North Carolina State. I still believe that the need to teach statistics to so many different disciplines and consult on such a variety of research projects justifies the development of an administrative unit which can serve all units of the university equitably.

My final service at North Carolina State was on the Faculty Senate. I was vice-chairman in 1965–66 and due to become chairman in 1966–67. During my stint as senator, I had many sessions with fellow senators at other North Carolina universities helping to strengthen the fringe benefits for faculty members and then dealing with problems brought on by a speaker-ban law which curtailed universities in their efforts to provide students with all points of view on controversial matters.

I should not conclude this résumé of activities in Raleigh without mentioning my political activities in 1964 as a member of the Executive Committee in North Carolina for Scientists, Engineers and Physicians for Lyndon Johnson and Hubert Humphrey. Nothing has so infuriated me as to realize we had fought so strongly to prevent expansion of the Vietnam war and then see that expansion actually take place.

8. A New Adventure in Kentucky

In 1966, I took leave from North Carolina State to spend another year with my old friend, Carl Kossack, this time in the Statistics Department at the University of Georgia. While there, I had two offers—one to become the Dean of the Graduate School at an eastern university and the other to become Chairman of a newly created Statistics Department at the University of Kentucky. Because the deanship would have resulted in my severing most of my connections with research in statistics, I decided to accept the Kentucky offer.

Leaving Raleigh in 1967 was the most excruciating experience Mary, Kathy, Bill and I have faced. Bill had enjoyed his year at Athens, but he had expected to return to Raleigh for high school. I realized that we were asking an awful lot of him to become acquainted with students in different schools in three successive years. His experience turned out to be even worse, because we moved to a different section of Lexington in 1968; hence, he had different sets of schoolmates in four successive years. The transition was not so difficult for Kathy, who completed high school at a junior college in Raleigh and went to college, first at my old Alma Mater, DePauw, and subsequently in Economics at the University of Kentucky. She returned to

Raleigh for graduate work and delighted all of us by obtaining her Ph.D. in 1978; she is now married and working as an Assistant Professor of Economics at Vanderbilt. Bill completed an Honors program in English at Kentucky. After flirting with graduate work in English, he decided that the job market was too limited in that area; he became involved in Kentucky politics, and is now working for the Kentucky Department of Human Resources while pursuing an MBA program at the University of Kentucky.

My wife, Mary, had to leave her home town to start life anew, but she has managed remarkably well. She was an invaluable aid in helping me recruit faculty, entertain visitors and promote a graduate program. She has found a secure niche in the university women's club, PEO, a local homemaker's group, and a developing interdenominational church south of Lexington.

The Statistics Department was established to provide service teaching and consulting for the university and to develop a broadly based graduate program in both statistics and operations research. The University of Kentucky had become a vigorous academic institution under leadership of President John Oswald; the College of Arts and Science (in which we were housed), along with many other colleges, had energetic deans, all of whom welcomed the development of a strong Statistics program. In particular, we developed a joint-appointee program with the three largest departments in the College of Agriculture.

I remained as chairman for 12 years, an unusually long tenure for Arts and Sciences, which prefers chairmen to serve no longer than two terms (8 years). In 1979–80, I remained at the University of Kentucky as a Professor in Statistics but in 1980–81 I switched my allegiance to the College of Agriculture as Assistant to the Dean for Statistical Services; the latter role enables me to handle consulting for departments not covered by our three joint appointees. By 1979, the number of permanent positions in statistics was increased to 17 (13 full-time and 4 joint appointees). We had assumed responsibility for teaching almost all the first-course undergraduate and graduate service courses in statistics on the campus. An excellent consulting program had been developed in the College of Agriculture and to some extent in Education.

I directed the graduate program during all but three years of this period; we managed to obtain 45 graduate majors in 1973–74 but the number has dropped to about 30 today. We have granted 23 Ph.D. and over 40 MS degrees; my personal contribution was to have directed or codirected 6 Ph.D. dissertations. It is becoming quite difficult to secure graduate students and even more difficult to persuade them to continue to the doctorate after completing the MS. There are many reasons for this: fewer undergraduates in mathematics from whom we have secured most of our graduate students; the attraction of computer science for those with a mathematical bent; the excellent industrial and government positions which do not require the Ph.D. One solution to the problem has been to recruit more foreign graduate students, but I feel that drastic methods are needed to redirect our statistical graduate programs. We must seek undergraduates in the biological and physical sciences and then offer them a graduate program which emphasizes statistics for the experimenter; we in the USA have allowed the theoretical statistician to dominate our statistical programs excessively to the long-term detriment of the profession.

It was an exciting period in my life, because I met so many excellent research workers and administrators who were interested in making Kentucky an outstanding university. We were able to obtain a good computing facility to sustain our teaching, consulting and research efforts, and I expect the statistics program to continue to increase its influence on the campus. As I look back over these developmental years, I am impressed by several salient points:

(i) It is difficult to develop a single broadly based statistics academic program. If one acquires a number of theoretical statisticians and probabilists, they may try to redirect the program towards more pure mathematics, probability and statistical theory. The joint appointees become unhappy, as they feel they are often regarded as pseudo-statisticians.

(ii) A chairmanship often does not carry the power needed to surmount the above pressures.

(iii) Administrative officials are often replaced by successors who have different priorities.

9. Some Other Activities: In Conclusion

During this period, I managed to attend the IBC–ISI meetings in Australia (1967) and the 1969 dedication of the Computing Center at the University of Cairo [10]. In 1970, the four Andersons and Larry Nelson joined Gertrude and two others on a VW-tour of the British Isles and West Germany en route to the IBC in Hanover; it included visits to the Passion Play in Oberammergau, many castles, beautiful countryside, and animated discussions. After meetings, the Andersons spent a week of sightseeing, attending plays and concerts and meeting old friends in London. (Details of the trip are available upon request!)

In 1974, Mary and I joined four other friends on a tour of Vienna, Budapest, Belgrade and Sofia *en route* to the IBC in Romania; Larry Nelson met us in Istanbul where we spent three hilarious days. The lack of freedom of association was striking in Sofia, and quite evident throughout the countries behind the Iron Curtain. We all breathed a sigh of relief when we landed in Zürich on our return flight. A striking feature of the IBC meeting was the session I chaired which was dedicated to Professor Snedecor; it was at this session that Larry made our first formal presentation of the linear-plateau models.

Our next international jaunt was to India in 1977, where I served as Visiting Professor at the Indian Statistical Institute in New Delhi. Many people doubted the wisdom of taking my sabbatical in India, but Mary and I feel that we should find out how other people live. We had a glimpse of many aspects of Indian life; we lived in a faculty apartment within the Institute compound, of which the walls and other buildings were being built while we were there. Across the street was a traditional Indian village beside a new development of apartments for tenants who could afford automobiles and television sets; a block away was a luxurious five-star hotel. Within walking distance were the remains of the first city of Delhi (old Delhi is the seventh). Then we could ride by crammed buses, three-wheel or regular taxis or by Institute cars. Despite this and because of the low-calorie diet, I lost over 20 pounds, a loss most of which I have managed to sustain since returning to Lexington.

I taught six students a course in Experimental Design, but I am afraid they learned little because they were taking seven other courses at the same time. We became quite friendly with the students and other visitors to the Institute. We appreciated the efforts of Professors Rao and Mitra and their wives to make our life enjoyable while there. The stay ended with the ISI meeting held in New Delhi and the local IMS meeting, where I presented a paper. I should mention that we spent an exciting week in London and three delightful weeks in Umeå, Sweden, with the Kulldorffs, prior to going to Delhi.

Our most recent international trips have been to the IBC in Brazil in 1979 and to Mexico City after the ASA meeting in Houston in 1980. As a part of the Brazilian trip, we toured Peru, where we visited the old Inca cities; I presented more linear-plateau material at the IBC.

For the past three years, I have been heavily involved with Section U of the American Association for the Advancement of Science (AAAS), of which I was chairman in 1979. In connection with the AAAS, I attended meetings in Houston, San Francisco and Toronto. An important assignment has been with the ASA Committee on Statistics and the Environment, of which I became chairman in 1980. We are arranging for improvements in the use of statistics in several of our federal agencies concerned with the environment, especially with the Environmental Protection Agency (EPA). I was also on the ASA's Census Advisory Committee for six years (chairman in 1977), a very important committee to advise the census on all aspects of its operations; of special importance was the pervasive "undercount" problem.

Since coming to Kentucky, I have found time to consult with a number of drug companies on clinical trials and have helped former students of mine with the International Mathematical and Statistical Libraries (IMSL), a computer programming company in Houston. I have also worked with Will Dixon in improving the BMDP computer programs at the UCLA Health Science Computing Facility. Then in 1978, Harlley McKean, Dennis

Haack and I established our own private consulting company, Statistical Consultants of Lexington, which we hope to expand. In my new position in the College of Agriculture, I have no teaching responsibilities but will work with an occasional statistics graduate student; I am also working with several research workers in the Medical Center on a proposal to study the impact of aging on memory loss. This seems to be an appropriate research area for someone in his late sixties.

I cannot end this autobiography without a few words of appreciation for the help rendered me, Mary and my family by our departed associate, former boss, friend and confidant, Gertrude Cox. She and I disagreed on many things, but on fundamental issues of where statistics belongs in the real world, we were in complete agreement. As for my family, Gertrude was always a staunch supporter. Her contributions to statistics and the statistical community are perhaps not fully recognized; if these few words can add to her recognition, I shall be content.

Publications and References

[1] Anderson, R. L. (1942) Distribution of serial correlation coefficient. *Ann. Math. Statist.* **13**, 1–12.

[2] Anderson, R. L. (1947) Use of variance components in the analysis of hog prices in two markets. *J. Amer. Statist. Assoc.* **42**, 612–634.

[3] Anderson, R. L. (1954) Components of variance and mixed models. *Qual. Control Convention Papers, Eighth Annual Convention Amer. Soc. Qual. Control*, 633–645.

[4] Anderson, R. L. (1956) A comparison of discrete and continuous models in agricultural production analysis. *Methodological Procedures in the Economic Analysis of Fertilizer Use Data*, ed. E. L. Baum et al., Iowa State College Press, Ames, 39–61.

[5] Anderson, R. L. (1957) Some statistical problems in the analysis of fertilizer response data. *Econ. and Tech. Analysis of Fertilizer Innovation and Resource Use*, ed. E. L. Baum et al., Iowa State College Press, Ames, 187–206.

[6] Anderson, R. L. (1960) Uses of variance component analysis in the interpretation of biological experiments. *Bull. Internat. Statist. Inst.* **37** (3), 71–90.

[7] Anderson, R. L. (1961) Designs for estimating variance components. Proc. Seventh Conf. Design for Experiments in Army Research Development and Testing, 781–823. Fort Monmouth, New Jersey. Institute of Statistics Mimeo Series 310.

[8] Anderson, R. L. (1965) Non-Balanced experimental designs for estimating variance components. Report of Seminar on Sampling Bulk Materials, U.S.–Japan Cooperative Science Program, Tokyo, Japan. Institute of Statistics Mimeo Series 452.

[9] Anderson, R. L. (1967) Non-balanced designs to estimate variance components. *Proc. Sixth Internat. Biometric Conf.* Sydney, Australia **5**, 117–150.

[10] Anderson, R. L. (1969) Use of computers in statistical teaching, consulting and research. *Proc. Inaugural Conf. Scientific Computation Center, Cairo*, 654–669.

[11] Anderson, R. L. (1975) Designs and estimators for variance components. Chapter 1 of *Statistical Design and Linear Models*, ed. J. N. Srivastava. North-Holland, Amsterdam.

[12] Anderson, R. L. (1980) William Gemmell Cochran, 1909–1980. A personal tribute. *Biometrics* **36**, 574–578.

[13] ANDERSON, R. L., ALLEN, D. M. and CADY, F. B. (1972) Selection of predictor variables in linear multiple regression. Chapter 1 of *Statistical Papers in Honor of George W. Snedecor*, ed. T. A. Bancroft. Iowa State University Press, Ames.

[14] ANDERSON, R. L. and ANDERSON, T. W. (1950) Distribution of the circular serial correlation coefficient for residuals from a fitted Fourier series. *Ann. Math. Statist.* **21**, 59–81.

[15] ANDERSON, R. L. and BANCROFT, T. A. (1952) *Statistical Theory in Research*. McGraw-Hill, New York.

[16] ANDERSON, R. L. and BATTISTE, E. L. (1975) The use of prior information in linear regression analysis. *Commun. Statist.* **4**, 497–517.

[17] ANDERSON, R. L. and CRUMP, P. P. (1967) Comparison of designs and estimation procedures for estimating parameters in a two-stage nested process. *Technometrics* **9**, 499–516.

[18] ANDERSON, R. L. and NELSON, L. A. (1971) Some problems in the estimation of single nutrient response functions. *Bull. Internat. Statist. Inst.* **44** (1), 203–222.

[19] ANDERSON, R. L. and NELSON, L. A. (1975) A family of models involving intersecting straight lines and concomitant experimental designs useful in evaluating response to fertilizer nutrients. *Biometrics* **31**, 303–318.

[20] BUSH, N. and ANDERSON, R. L. (1963) A comparison of three different procedures for estimating variance components. *Technometrics* **5**, 421–440.

[21] GREGORY, W. C. and ANDERSON, R. L. (1975) Design procedures and use of prior information in the estimation of parameters of a non-linear model. *Commun. Statist.* **4**, 483–496.

[22] KUSSMAUL, K. and ANDERSON, R. L. (1967) Estimation of variance components in two-stage nested designs with composite samples. *Technometrics* **9**, 373–389.

[23] MOTE, V. L. and ANDERSON, R. L. (1965) An investigation of the effect of misclassification on the properties of χ^2 tests in the analysis of categorical data. *Biometrika* **52**, 95–109.

[24] MUSE, H. D. and ANDERSON, R. L. (1978) Comparison of designs to estimate variance components in a two-way classification model. *Technometrics* **20**, 159–166.

[25] SAHAI, H. and ANDERSON, R. L. (1973) Confidence regions for variance ratios of random models for balanced data. *J. Amer. Statist. Assoc.* **68**, 351–952.

[26] STROUP, W. W., EVANS, J. W. and ANDERSON, R. L. (1980) Maximum likelihood estimation of variance components in a completely random BIB design. *Commun. Statist.* **9**, 727–756.

D. J. Finney

David John Finney was born at Latchford, Warrington, Cheshire in 1917. He was educated at Lymm and Manchester Grammar Schools, and in 1934 went on to read Mathematics and Statistics at Cambridge University. He studied for a year with Sir Ronald Fisher at the Galton Laboratory, London, and was then appointed Assistant Statistician at Rothamsted Experimental Station in 1939. In 1945 he took up a lectureship at the University of Oxford, and in 1954 became Reader at the University of Aberdeen. In 1963 he was appointed Professor of Statistics at Aberdeen, but moved to the University of Edinburgh three years later in 1966. Since 1954 he has also been director of the Agricultural Research Council Unit of Statistics.

Professor Finney has been closely involved with research on the biological applications of statistics, and on experimental design. He has served as consultant to the FAO, the WHO, the Cotton Research Corporation, and the UK Committee on Safety of Medicines. He has also been Chairman of the UK Computer Board for Universities and Research Councils between 1970 and 1974. He has taken an active role in statistical affairs both internationally and in the UK. He was elected President of the Biometric Society in 1964, and of the Royal Statistical Society in 1973; he is a member of the International Statistical Institute. Among his many honours are the Weldon Memorial Prize, and the Paul Martini Prize (for clinico-pharmacological methodology), honorary doctorates from Gembloux, Belgium, City University, London, and Heriot-Watt University, Edinburgh, and the CBE.

Professor Finney is the author of eight books and numerous papers on statistical topics. He is married with one son and two daughters. His recreations are travel, music and the three Rs.

A Numerate Life

D. J. Finney

In the development of an individual, nature and nurture both play their parts, though for those who have had the good fortune to be brought up within a normal and happy family the two are not easily disentangled. As I look back at my own life, I see both innate qualities and early environment as having recognizable effects. I also know the dangers of subjective judgement: possibly more is due to the good luck of having been in the right place at the right time on several occasions.

1. Childhood

I was born in 1917, the eldest of three boys. My father, son of a teacher who presumably had little sympathy with anything that might be called higher education, began work in his mid-teens on the accounting staff of an iron and steel company in Warrington. He placed great value on loyalty to his employers, to an extent that would seem strange today and that was never matched by the care of an employer for his employees. He followed his original firm through mergers until his retirement as Chief Accountant to the Lancashire Steel Corporation. Had the educational opportunity occurred, his excellent sense of number and his basic mathematical understanding could have made him successful in mathematics. My mother came of a long line of Cheshire farmers. Though her formal education was limited, she was a skilful teacher. I gained from the early years of being the only child, whom she had time to teach the three R's; I must have begun to read and write at about four, and certainly by five reading was becoming a pleasure. I also remember with affection her father, a strict disciplinarian who expected instant obedience from his grandsons but whose principles, wisdom, and sense of humour were an important part of my childhood environment.

One early incident deserves recording, my first careful planning of a scientific experiment. I have been told by physiologists that this is impossible, but I know it happened: perhaps a small child is less constrained by some of the reflexes that later govern our actions. I was about three years old. I stood on a rug in our kitchen. I realized that now not only could I stand firmly on two feet but that I also had a tolerable balance on one foot alone. Could this be generalized? I clearly remember the process of thought that led me to plan raising the remaining foot. I did not then know the word "extrapolation," but from approximately one second after the ensuing disaster, I have distrusted the idea profoundly. I shall always maintain that the incident was a well-reasoned attempt by a small child to add to his understanding of the world around him. Certainly the result impressed me strongly, a convincing demonstration that in some circumstances an unreplicated experiment is an adequate basis for inference. Those were days when, especially in a small provincial town, families still largely made their own entertainments. I was soon an omnivorous reader; puzzles of all kinds, card games, and games of skill ranging from noughts and crosses to chess did much to develop verbal and numerical ability. Especially during the years of economic depression, when father worked long hours under the daily fear that his job would disappear, mother held the family together, and ensured that my brothers and I have happy memories of a childhood that was rich in variety at the time though doubtless it would seem dull to today's teenagers. The wide experience of music and the visual arts that now mean much to me, however, were pleasures that had to come much later.

I spent five formative years at the Lymm Grammar School, a small coeducational school with high standards. W. B. S. Hawkins, my headmaster, was a fiery little man with strong views on discipline, from whom I began to learn something of the principles of scholarship. Thanks to mother's early efforts, I was soon in classes with children two to three years older. Today's educational dogma would condemn this separation from an age group as bad for social and emotional developments; I remain grateful that I was not denied the mental challenge that I needed. Because I was younger and physically less powerful than my classmates, I avoided wasting excessive time on sports that (except for the numerical intricacies of cricket) never made any appeal to me. In most subjects I could perform well, but "Art" was an exception: our somewhat unimaginative instructress tried to persuade me of the principles of perspective, but at the age of 11 I could not reconcile my knowledge that parallel lines did not meet with her insistence on making them do so. I still recall my first research excitement, when a trivial generalization of a formula from my analytical geometry text brought realization that I could handle a problem with more vision than could my teacher of the time.

In 1931, Hawkins arranged my transfer to the Manchester Grammar School, for which generous sacrifice of one of his best pupils I shall always be grateful. Mathematics at Manchester Grammar School was then in the charge of R. C. Chevalier, a true eccentric but a great teacher, to whose sense of rigour and appreciation of style in presentation I owe much. At that period, specialization in mathematical education in England was extreme; during my last four years at school, serious study was almost restricted to mathematics and physics, and my subsequent undergraduate years were spent solely on mathematics. Educationalists today would disapprove, but my only regret is that I had no formal teaching in biology.

2. Cambridge

At 17, life broadened for me immensely when I entered Clare College, Cambridge. Unsophisticated, naive, and narrow in outlook, I entered a society more diverse in interests, backgrounds, and abilities than anything I had known before. I revelled in the opportunity of concentrating all mental effort on the many aspects of mathematics, taught by some of the leading mathematicians of the day, and having access to an excellent library. I fully accepted the Hardy view of the supremacy of pure mathematics. Living in what I still find the most beautiful of urban environments taught me that appreciation of beauty was not restricted to creative artists. Now that participation in sport was a free choice, I learned to enjoy tennis and squash, and gained much from the camaraderie of rowing: fortunately academic success was not dependent upon quality of performance here!

In 1936, a seeming disaster proved a piece of great good fortune. Examination results convinced me that I was at best a good second-class mathematician. Daydreams of a brilliant academic career disappeared, neither actuarial work nor school teaching had the slightest appeal, and other ways of earning a living from mathematical skill were few. Then came my only long spell of illness, first an extended rest because of a suspected lung weakness and next a very unpleasant attack of typhoid fever. In consequence, I missed a term in Cambridge and found my choice of optional special subjects much restricted. I had little idea what "Statistics" might be, but it seemed worth a trial.

John Wishart never tired of commenting on the number of the rising generation of statisticians in the 1940s who had been his students. Though he never established a school of statistical research, undoubtedly his teaching was an important influence on many. His presentation lacked the rigour that theorists would demand today, yet was perhaps unduly formal in its approach to practical problems. He did insist on associating numerical exercises with mathematical theory, and he evidently succeeded in giving young mathematicians an appetite for this expanding subject. After his course, my thoughts on a career took a new turn; grants for graduate study were then practically non-existent, but my father generously financed me for a further year to work with Wishart. Derrick Lawley and I, exact contemporaries, shared an introduction to research in an unstructured manner that I believe far better for statisticians than the narrow road of a Ph.D. We learned to read widely in journals, seeking problems for further exploration in the publications of others. We looked at data for people who came to consult Wishart; in particular, we undertook extensive analyses on sheep growth and nutrition for a South African visitor to Cambridge. We did a little tutoring. I remember with amusement a scientist from another field who asked me for help in understanding the *t*-test, and subsequently wrote one of the worst of introductory statistical texts. Early in 1938, I responded to an advertisement for scholarships offered by the Ministry of Agriculture. To my surprise I was called for interview; this intimidating experience, the sole occasion on which I have had to face an interviewing panel, turned out well and I was awarded a two-year scholarship to work under R. A. Fisher.

3. Galton Laboratory, London

I knew Fisher only from one uncomfortable interview in 1937 when, after a brief preliminary talk, he put me in a room for an hour to develop original thoughts on the "Problem of the Nile" as expounded in his recent paper; needless to say I was devoid of ideas, and a year later I fervently hoped that he had forgotten this meeting. Though Wishart's teaching was strongly Fisherian, he had talked little about Fisher as a man. My picture was compounded of the fact of authorship of innumerable papers and books and the folklore that saw him as a strange genius of uncertain temper and little tolerance for others. Though I had to learn that some topics needed to be approached delicately, he was to me always a kind and helpful teacher and increasingly a valued friend.

The ensuing year was invigorating. Fisher himself overflowed with original ideas. His staff and research students included A. C. Fabergé, Horace Norton, V. G. Panse, Rob Race, Tony Stevens, G. L. Taylor, Helen Turner, and Evan Williams, and distinguished statisticians and geneticists of the day were frequent visitors. Recognizing my biological ignorance, Fisher urged on me a small investigation with his mouse stock. He had a strain carrying a gene that in heterozygotes produced a marked shortening of the tail and in homozygotes was lethal. Fabergé, ever ingenious with apparatus, devised a box with a slit in one end to facilitate the measurement of tail length. Sadie North and I then embarked upon a programme of examining the tail length of heterozygotes, and selectively breeding from the longer-tailed in order to see the effects on the homozygous lethality. We were indeed able to produce somewhat longer-lived homozygotes, but the project was more important to me in demonstrating biological variability and its measurement than it was for the study of mouse genetics. I also gained from participation in Fisher's discussions with Panse on cotton breeding, not least because I often had to assist Panse later in deciphering scarcely legible notes or interpolating missing steps in a mathematical argument through which Fisher had passed too speedily.

With typical generosity, Fisher suggested that I should generalize his work on estimating genetic linkage from records of human families; I developed his ideas in my first major piece of research, a series of papers that were extensively used by many people. (Today a computer and a general maximum likelihood program will do the job more expeditiously than these specialized scoring techniques.) My own reading led me to study the distribution, under the usual normal assumptions, of the largest of a set of variance ratios based on independent mean squares in their numerators but sharing the same denominator. Having enjoyed some entertaining mathematics, obtained a general result, and tabulated for the simpler cases, I showed the work to Fisher; my pride took a fall when he found it of no interest and not worth publication. Two or three years later, just after he had accepted for publication some related work by W. G. Cochran, I dusted off my old manuscript and submitted it to him for the *Annals of Eugenics*: he expressed delight in a very nice paper! In retrospect, I think his first judgement was perhaps the correct one.

4. Rothamsted

Early in 1939, Fisher suggested to Frank Yates that I might be a suitable replacement for W. G. Cochran, who was about to leave Rothamsted for a new position in Ames, Iowa. He offered me the job, I suspect less because of any evident abilities than because there could then have been no one else of my age in the country with anything approaching a suitable training. I hesitated about giving up my scholarship and the possibility of a Ph.D., but this seemed precisely the kind of opening for which the scholarship was intended to fit me and nothing comparable was likely to be available a year later. Around midsummer 1939 I moved to Rothamsted for a short overlap with Cochran; he left, Yates went on a three-week holiday, and I found myself acting as "the" Rothamsted statistician.

The six years of war gave new impetus to agricultural research. That we who worked at Rothamsted were privileged by safety, comfort, and a constructive pattern of life is true, but equally truly we all worked hard and collectively achieved much towards keeping the country fed. The need for statistical evaluation of results from experiments and for quantitative survey of agricultural requirements was increasingly recognized. Yates played an outstanding role in agricultural research policy and fought hard for the acceptance of new ideas. My part was mostly to maintain the general statistical activity of the station while Yates pursued special projects at a higher level, though on some things we worked closely together. From him I learned much about experimental design, both in its combinatorial aspects and on broader questions of devising a design optimal for the study of certain questions. A legacy from Cochran was statistical support for the Insecticides Department. I was soon deep in the mysteries of quantal responses and probits, finding a challenge in another class of maximum likelihood problem. Those who today have ready access to computers and to good optimization subroutines may not realize how often theoretically desirable estimation by maximum likelihood used to be obstructed by computations that were tedious to the statistician and almost prohibitive to a biologist.

Crop sampling for estimating pest infestation presented many statistical problems. One major group arose primarily because much old grassland was being broken for arable cropping, a practice favouring substantial wireworm damage to the first few crops. Yates devised a simple sampling scheme for estimating wireworm populations in a field; I worked with the Ministry of Agriculture advisory entomologists in applying this throughout England and Wales, with a view to identifying fields that called for resistant crops or other special precautions. Apart from studies of size of sampling unit and of the relation between population density and sampling variance, the formal statistical content was slight. However, this co-operative exercise, together with varied work on the design and analysis of field experiments for Rothamsted and other centres and a steady flow of problems in insecticide testing, were invaluable in teaching me how closely a biometrician must work with biologists if he is to understand the real questions and to avoid introducing unnecessarily abstract theory.

Opportunities for development of statistical methodology were plentiful. The wireworm work had raised questions of how far final crop yields suffered through random elimination of seedlings at an early stage. We wanted to test this experimentally, but could not justify special experiment. Yates suggested that, if in a single replication of a 2^5 fertilizer experiment on sugarbeet an additional factor at two or even four levels were to be inserted on a small part of each plot, the interesting effects might be confounded with high-order fertilizer interactions. He left the details to me, and so I had the enjoyment of being the first to elaborate the elegance and the practical utility of fractional replication. I was still working on genetical problems. My early publications on probits and insecticide tests began to bring me other contacts in biological assay, notably with two chemists in the pharmaceutical industry, A. L. Bacharach of Glaxo and Eric Wood of Virol. The days of wireworms and analysis of field experiments left little time for this increasing range of problems from outside agriculture, but the weekends and nights spent at Rothamsted on guard against incendiary attacks gave opportunities for intensive work and a steady output of research publications.

5. Oxford

Fisher gave me a strong hint that the University of Oxford had me in mind for a new post for a statistician. When the university advertised the Lectureship in the Design and Analysis of Scientific Experiment, I applied.

Some weeks later I received a letter to say that I had been appointed, without interview and without enquiry as to whether I wished to accept. Very conscious of my debt to Frank Yates, I moved to Oxford in the autumn of 1945. At first, I operated my new unit not only without staff but without furniture or equipment. For several days, until David Lack of the Edward Gray Institute lent me a chair and small table, I sat on the staircase writing a paper. The new lectureship had been established to give statistical help to biology and medicine, though my appointment expressed no conditions on what I should do. I rapidly made contacts, obtained introductions to various departments, and let it be known that I was willing to be a consultant, perhaps even to undertake some analyses. I began a course of introductory lectures for biologists, which in the first year or two attracted a number of students and junior research workers who subsequently made their names in many fields.

With P. H. Leslie in the Bureau of Animal Population and David Kendall steadily moving his mathematical interests towards statistics, both of whom became close friends, Oxford was a good place for extending statistical activities. An early step was to start a Diploma in Statistics, at a time when scarcely any British university offered an explicit qualification. The first recruit to my staff proved an excellent choice; Michael Sampford was to remain my colleague for more than 20 years. My research was increasingly concerned with biological assay. Of the many papers I then published, "The Principles of Biological Assay" in the *Journal of the Royal Statistical Society* was perhaps the most important; it was the first major attempt to give logical structure to the whole group of problems. My book *Probit Analysis* aimed to be different from most texts at the time, a thorough study of one family of special techniques rather than a general account of statistical theory and practice. Although I now wish it had been less narrow in concept, it proved very successful in helping scientists to use what then seemed difficult methods of statistical analysis, and its popularity has been sustained for 35 years. I gave more practical outlet to my bioassay interests by contributing an elementary chapter to Joshua Burn's *Biological Standardization*, a book much used by pharmacologists. My first visit to the USA, to lecture to a Chapel Hill Summer School in 1949, prompted me to expand my ideas on biological assay as a special field of statistics: the lectures later grew into my favourite book.

Three other lines of work from my time at Oxford seem worth recalling. I undertook sampling studies with particular reference to the estimation of timber in standing forests. I endeavoured to make a clear presentation of the case for random sampling, and was able to display a genuine but somewhat mysterious example in which systematic sampling could be seriously wrong. Fascinated by the peculiarities of Latin squares, I considered various less than complete orthogonal partitions, especially of the 6×6 squares. Interesting combinatorial properties emerged, and pointed to further work that could not be attempted with the facilities of the time. At the instigation of A. C. Fabergé, I attempted to study equilibrium conditions for species in which a genetically controlled polymorphism inhibits crossing between individuals of the same phenotype. Lengthy calculations for *Oxalis rosea* and for several imaginary genetic specifications suggested a widespread but not universal rule that equilibrium would entail equal frequencies for the different phenotypes. I was able to make some theoretical progress but nothing of sufficient generality. Recently, I have returned to these last two topics, largely as a result of having my own microcomputer. With its aid and subsequent transfer of programs to a larger computer, I have been able almost to complete the wide range of enumeration problems concerning orthogonalities in the 6×6 Latin squares. My 1952 paper on polymorphic equilibria has from time to time drawn others (especially Spieth in the USA and Heuch in Norway) to work on this intractable group of problems. My desk-top micro is now busy helping me in further explorations.

A major consequence of my summer in the USA was my marriage in 1950. This is not the place for presenting personal as distinct from scientific life, but the two cannot be separated completely. For more than 40 years, statistics has been an addiction that had to take a high priority and that could never be entirely set aside. I am by nature impatient, and when I am making progress with a problem, preparing a book, writing a computer program, or anything of the kind I am not easily distracted by family and social matters. Only a very patient wife not merely tolerates such behaviour but creates an atmosphere in which it seems natural. Without Betty's long-suffering help, encouragement when difficulties seemed overwhelming, and creation of a happy home, any achievements of the last 30 years would have been very different. Though she and my three children might strenuously deny much comprehension of the details of my professional life, their contributions to it are immeasurable.

In 1952, Frank Yates and I were asked by the UN Food and Agriculture Organization (FAO) to assist the statisticians of the Indian Council for Agricultural Research. We agreed that I should spend a full year in New Delhi and that Frank would come for one or two shorter periods. India had only recently achieved independence, and it was with some trepidation that I took my wife and infant daughter. We had nothing but kindness from the Indian people, and my chief worries were the bureaucratic complexities of FAO. The opportunities of seeing many parts of this remarkable country gave Betty and me memories that we still treasure. If there was danger that the atmosphere of Oxford was diverting my statistical interest from reality, India was a great corrective.

Frank and I regarded one major task as being to pull together results of fertilizer experiments on the major crops, in the hope of extracting general guidance on fertilizer use. In this we followed what he and E. M. Crowther had done for British crops 10 years earlier. It proved an excellent way of bringing some young Indian statisticians into closer touch with experimentation and also gave valuable results, not least a clear indication of the value of potash on many soils that previously (without reference to evidence) had been asserted to be unresponsive to potassic fertilizers. I concerned myself much with the practice of experimental design, endeavouring to propagate the understanding that combinatorial aspects of the design of the experiment should correspond with the questions being asked rather than be taken from a textbook as the first pattern roughly agreeing with the size of the experiment intended. A survey of experimental designs then in use disclosed some monstrosities and showed the need for far more attention to planning the use of research resources. In addition to undertaking specific tasks, Frank and I were asked to advise on the whole structure of agricultural research statistics in India; suggestions for reorganization, which were in part implemented, were an important component of our eventual report.

6. Aberdeen

Soon after my return from India, the Agricultural Research Council (ARC) and the University of Aberdeen invited me to lead a cooperative development in statistics. The University would set up a Department of Statistics and the ARC would add to it a unit charged with research and consultative responsibilities. I have found that occasional changes in working environment are stimulating to thought, and I recognized that further advance in biometry in Oxford required someone to give it new direction. (Characteristically, my father asked whether I was treating the University of Oxford fairly by withdrawing my services.) My new team came together in the autumn of 1954 with plentiful opportunities ahead. A good omen and pleasant surprise for me personally was my almost simultaneous election to the Royal Society and to the Royal Society of Edinburgh, early in 1955. Within the university, the need for giving basic understanding of statistics to students and providing statistical support and collaboration for a wide variety of research programmes was evident throughout the faculties of science, medicine, and social science. We also began to train those who would become professional statisticians, through undergraduate courses within the mathematics degree and a postgraduate diploma that eventually became the M.Sc. We had a small but steady flow of good students: I have never regretted my insistence that carefully selected graduates from disciplines other than mathematics should not be barred from graduate study in statistics.

The new ARC unit soon found its place in Scottish agricultural research. It could offer help to about a dozen research institutes and agricultural colleges. In that pre-computer age, it was welcomed as an agency that could help with the analysis of experimental and other results; often, but not often enough, this led to involvement in the planning of further research and in many aspects of the interpretation of data. We were active in sample survey; in particular, we adapted a Rothamsted scheme for the survey of fertilizer use, and in association with the agricultural colleges conducted surveys of farms in several areas per year for 10 years. In the late 1950s, fall-out from tests of nuclear weapons rightly caused great concern. I was asked to help in monitoring contamination of food supplies, especially by following strontium-90 levels in milk. We devised a relatively simple sampling plan that allowed time trends and broad regional differences to be estimated with very satisfactory precision, making British information on this subject more trustworthy than that of any other country.

My personal research concentrated on the optimal planning of experiments in relation to the questions to be answered, a much broader topic than the purely combinatorial aspects of experimental design. A paper to the Royal Statistical Society in 1955 emphasized the wholeness of a statistician's responsibilities in agricultural experimentation. Subsequently I looked at an idealized version of the planning of successive stages of selection among many new varieties of a crop. Consideration of how hundreds or thousands of potential new varieties should be reduced to perhaps about five in a few years of selection raised an interesting spectrum of questions. Pursuing my belief that the explicit objective should be to optimize the performance of the varieties finally selected rather than to establish significance tests or to maintain some notion of fairness towards every variety, I was able to show convincingly that an optimal stable selection process was almost achieved by fairly simple rules; although near-optimality is insensitive to quite considerable modifications in the parameters of the process, gross departures may imply inefficient use of resources. My specification of the problem could not correspond perfectly with reality, but the results at least provide a frame of reference; I regard this work as embodying one of my few truly innovative ideas. A few years later, in a tentative manner, I applied similar methods to selection in the education of children, now necessarily introducing requirements of fairness to the individual child. Unfortunately, I completed this just at the time when suggestions of élitism in education were becoming subject to grave disapproval. I had never imagined that results from my crude specification would be directly applicable. What I had done was to recognize the need to specify objectives for an educational system, taking account of individual rights and community benefits, and then to seek a structure that would optimize approach to the objectives. The detail is of no lasting value, but perhaps one day the broad intention will be accepted in place of the assumption that policies fluctuating with the fashions of sociological theory will necessarily maximize benefit to the individuals and maintain educational advance from generation to generation.

Since taking over from him at Rothamsted, I had known Bill Cochran only through occasional meetings and letters, but in 1962 he suggested that I might join him at Harvard for a year. I needed the change and gladly accepted the arrangement of spending half my time in the Department of Statistics, half with Professor Dave Rutstein in the Department of Preventive Medicine. We all benefitted. Betty renewed contacts with family and friends, the children gained enormously from exposure to different schools and a different pattern of life, and I was able to begin new lines of work.

Frank Yates has enjoyed pointing out how I failed to see the coming importance of computers at the time when he was trying to tame the early machines. He is right, but possibly I have since atoned. I had been instrumental in getting the first computer for the University of Aberdeen, and in 1962 I was using it heavily to advance my selection calculations. Harvard had more advanced facilities, and having secured an allocation of a few hours of time I was able to generalize results I had been obtaining in Aberdeen. My first talk with Dave Rutstein took place when the thalidomide danger was becoming widely known. He suggested that we work together in examining how statistical approaches might give earlier warnings of drug dangers. I looked forward to further discussion of an interesting idea; Dave took off for several weeks in Europe, and I went ahead alone. I began with ambitious notions of widespread obligatory recording of all uses of therapeutic drugs and the subsequent history of patients. Reality crept in: compulsion was impossible and progress would have to depend upon less complete records spontaneously submitted. For any orthodox statistical analysis, the data would be highly suspect because of unknown selective influences. Yet even under such circumstances I could not believe that a series of reports of some serious adverse medical experience following administration of a specified drug was meaningless. Indeed, much former evidence that had led to the withdrawal or restriction of certain drugs was of this type. No one now seriously doubted that thalidomide early in pregnancy was dangerous; surely the danger could have been recognized sooner had there existed a systematic channel for reporting single instances of unexpected events.

I am still unsure whether this subject is truly "statistics," but it is certainly one where statistical habits of thought help to keep the mind clear and to develop a logical approach. I wrote a series of papers in which I advocated establishing registries for reporting of adverse events and suggested ways of using the ensuing files for monitoring drugs and obtaining signals of possible dangers; I think that at least two of these papers have been important in giving direction to subsequent developments in several countries. After returning to Aberdeen I soon found myself pulled into the work of the Committee on Safety of Drugs (now Committee on Safety of Medicines) and also to many special committees of the World Health Organization. I have found this a most rewarding activity. I have no doubt that it illustrates how an applied statistician can be a useful member of an operational group concerned with a very different discipline, provided that he has the ability to assimilate new ideas and to present a quantitative outlook cogently and clearly. Though I may now have done all that I can on this subject, I have no regrets at giving much time to it over 15 years.

7. Edinburgh

Once again a year abroad proved unsettling. Soon after returning to Aberdeen, I was invited to consider a return to Oxford, with a college fellowship and freedom to interest myself in any form of biological statistics that took my fancy. The prospect was appealing, but I realized that I was accustomed to leading a group with an obligation to give widespread service; I did not see myself as the right person to flourish alone, and the chances of financing staff to work with me were not good. I declined the offer, not without guilt in thinking of various friends who had worked at producing it. Not long afterwards, the Agricultural Research Council negotiated with the University of Edinburgh on the basis that my unit would be strategically better placed in Edinburgh and that the university needed to expand its statistical facilities. After much discussion, and consultation with senior staff who were affected, I agreed to accept a new chair and set up a department in the University of Edinburgh, bringing with me most of the unit staff.

Our operations in Edinburgh began in 1966, with an enlarged university department and opportunity to enlarge ARCUS (the ARC Unit of Statistics): a particular pleasure was to be joined by one of my earliest Cambridge friends, Derrick Lawley, as Reader in Statistics. In some respects the history of the next few years parallels that in Aberdeen. New courses had to be established, on a broader basis than before, but conflict remained between plans for giving a young biologist or social scientist what the statistician sees as a minimally adequate statistical foundation and timetable pressures that for many students limit the total exposure to statistics to 10 or 20 hours in four years. As I draw to the close of my teaching career, I remain broadly content with our professional teaching of mathematicians and those who will become practising statisticians, profoundly dissatisfied with what we do for students of the many other disciplines where understanding of statistical principles and a working knowledge of methodology are needed. Though we have the harder task, what we achieve compares poorly with what computer scientists succeed in putting across. To do better perhaps requires a radical rethinking of which I am incapable.

In Edinburgh, ARCUS is much more centrally placed relative to most of its clients, and I believe has become far more effective as a consultative and collaborative body. For many years its research performance was overshadowed by its service activity, but of late there has been marked change. This is primarily a personal story, and the main developments have not been directly in my own work; nevertheless, I must mention the significant part ARCUS now plays in research for the testing of new crop varieties, in quantitative genetics and animal breeding, and in sampling techniques for animal populations.

8. Experience and Outlook

I remain convinced that a university statistics department must be active in the application of statistics both for the benefit of other disciplines and in order to keep statisticians and their students firmly in touch with the true nature of their discipline. Unfortunately, to do this on the scale that is needed calls for teaching time in other disciplines where timetables are always under severe pressure, and also for staff who can find time to become deeply involved in the research projects of others. Even in periods of expansion, university finance has never been able to provide adequately for interdisciplinary support of research; it is scarcely surprising that a research group in any one science prefers staff or equipment of its own to expenditure on general statistical facilities. Thinking back over experiences in three universities, I realize that I have probably made the error of endeavouring to do something for everyone who sought co-operation instead of building a department with strength in one or two special fields of application. Greater concentration of effort could have created a firmer base for expansion.

My personal interest in the detection of adverse reactions to drugs has continued, with emphasis on using spontaneously submitted information as a means of detection or at least arousing suspicion even though any form of rigorous inference is impossible. I have increasingly turned back to work on biological assay. In a university whose Department of Pharmacology was successively led by Clark, Gaddum, and Perry, this might seem natural, but the stimulus to my own research has come almost entirely from elsewhere. A chance encounter with Egon Diczfalusy of the Karolinska Institute in Stockholm led to collaboration in a series of problems concerning hormone assays, the most interesting perhaps being the design and analysis of experiments for estimating the potency of anti-sera and of anti-anti-sera.

In the 1960s, advocates of the new radioimmunoassay techniques claimed them to be so accurate and so precise that need for statistical analysis was eliminated. Undoubtedly these and other immunoassays represent a great advance, especially because they can work with exceedingly small amounts of the materials to be assayed, an important consideration when assaying human sera for diagnostic purposes. Whatever may be true in specialist laboratories, in routine use immunoassays still display a variability in response that calls for statistical analysis. New forms of non-linear response curves must be handled, responses must be suitably weighted, and new problems of data handling arise. I have sought to bring radioimmunoassay within the general statistical framework of bioassay, while satisfying the special conditions of assay and the special needs of data processing. These assays frequently concern the treatment of individual patients, and statistical techniques that lose or distort an appreciable part of the information in the data cannot be tolerated. I think I have had some success, but progress involves struggles against vested interests as well as for statistical quality.

In 1966, I was asked to serve on the newly created Computer Board for the Universities and Research Councils, a somewhat surprising consequence of having helped the University of Aberdeen to formulate a strong case for a replacement computer. For the next nine years, I was active in trying to raise the standard of computer facilities in the British universities to a satisfactory level. When we began, many universities were without a computer, few had anything that could be considered adequate even for 1967, and there was neither co-ordination between universities nor a coherent policy for development. To work under the dynamic leadership of Brian Flowers, of course with the essential backing of reasonable earmarked government funds, was a stimulating and instructive experience. When after four years I took over from him as chairman, I had a smoothly running organization, a well-defined policy, and a great amount of goodwill from universities, many of which had been at first distrustful of this new form of interference with their independence. Hard work continued, but we were now a group of colleagues who could pull together; though I am clear that regular change in leadership is essential for such a body, the time when I had to hand over the chair seemed to come too soon.

A life spent in teaching and research brings neither public acclaim nor enduring visible monuments, but its satisfactions are not less great. I am not a very good teacher, nor do I greatly enjoy the process of teaching; as I stand in front of a class, I am never without the feeling that either its members are bored by what I have to say or that they find it far easier than I do. Nevertheless, even the more tedious aspects of teaching are amply rewarded when one learns that a student has been helped to success by one's efforts—whether this be a major professional advancement for a young statistician or a recognition by a biologist that enlightenment on his work has come through a simple statistical approach. My research has been concerned less with radically new ideas than with making rough places plain in statistical methodology. In several books and 200 papers, I have tried to extend the systematic use of established statistical principles and to put methods into readily applicable form. I have found enjoyment in doing, satisfaction in achievement.

I should be dishonest to pretend that more public honours have meant nothing. To be elected President of the Biometric Society (1964–65) and later of the Royal Statistical Society (1973–74) was a great joy, though I see such positions as opportunities for service and leadership rather than as pure honours. Honorary degrees from the Belgian State Faculty of Agriculture in 1967, from City University, London in 1976, and from Heriot-Watt University in 1981, the Weldon Medal of the University of Oxford, and the

Martini Prize were more personal. Above all, Her Majesty's appointment of me as a Commander of the British Empire in 1978 brought both personal pleasure and happiness that the work of my whole group over many years was recognized.

9. What Lies Ahead?

Dr. Gani suggested that each of these essays should end with some prediction of the future. I hope and believe that the future can have a continuing place for the two organizations that I shall leave behind, ARCUS and my present department. Within the areas that interest me, I see full and wise exploitation of the computer as the pre-eminent need. For most problems of experimentation, sampling, and general investigation in agriculture and biology, we have the statistical methods available, and the chief difficulty lies in making best use of them. We have still to come to terms with an age in which the cost of computation (in time or in cash) for work of this kind is almost negligible, but the choice between methods and the interpretation of many alternative analyses of the same data remains difficult. If we are not to mislead others and to be ourselves misled, we have to control computing excesses, not because they are costly but because they can confuse instead of elucidating. At one extreme, multivariate data—possibly with 100 or more variates from one experiment—challenge us to devise procedures that will encourage right interpretation and reduce subjective bias from a choice between many analyses. At the other, we must recognize that availability of computer packages encourages scientists to go directly to the computer without seeking statistical advice. To fight this trend, or to complain about the inadequacies of a popular package, is useless. For our profession today, one of the greatest responsibilities must be to produce improved packages that contain internal warnings against unwise choices, that suggest when aspects of the data call for professional statistical help, that produce good clear summaries of analyses, and that are at least as attractively easy to use as some of those packages that now offend us. We need software flexibly designed to meet statistical requirements, not statistical practice constrained by the views of software writers.

If statisticians experienced in biometric and other applications can see this as an intrinsically fascinating challenge over the next 20 years, in preference to the invention of more minor variations of method, our duty to science and technology will be well done.

Tosio Kitagawa

Professor Tosio Kitagawa was born at Otaru City, Hokkaido, Japan on 3 October 1909. He completed his high school education in Sendai City, and then decided to study mathematics at the University of Tokyo. He graduated there in 1934, and went on to do research on functional equations for which he was awarded a Doctorate of Science in 1939.

He began his academic career in the Department of Mathematics at Osaka University in 1934, and in 1939 was appointed Assistant Professor in the Department of Mathematics at Kyushu University. He was promoted to a Professorship of Mathematical Statistics in 1943, became Dean of the Faculty of Science at Kyushu in 1968–9 and Director of the Research Institute of Fundamental Information Science in 1968, a post which he held until his retirement in 1973. This was the beginning of a second career as Director of Fujitsu's International Institute for Advanced Study of Social Information Science, a position which he currently holds.

Professor Kitagawa has had considerable influence on the development of statistics and information science in Japan. He was awarded the Deming Prize on Statistical Quality Control by the Japanese Union of Scientists and Engineers in 1953, and several honours, including the Honorary Medal of the Japanese Government in 1980 for his scientific achievements. He is a member of the International Statistical Institute, and was President of the Information Processing Society of Japan in 1975–7.

Professor Kitagawa was married in 1936 and has three daughters and a son. He has travelled widely and made scientific contacts in Europe, the USA, India and Australia, which he actively maintains.

From Cosmos to Chaos: My Scientific Career in Mathematics, Statistics, and Informatics

Tosio Kitagawa

1. Introduction

I was born on 3 October 1909 in Otaru City, Hokkaido, Japan. I was educated there until middle school; I received individual tuition in English, mathematics and Japanese calligraphy. During my three years of high school (Niko) in Sendai City, I was very interested in philosophical meditation, and spent a great deal of my time reading difficult treatises on philosophy in the quiet academic atmosphere of the city. My last year of high school was spent in a Buddhist monastery intended for high-school boys. These experiences appear to me to have constituted the background of my career. After some hesitation and doubt, I determined to choose mathematics as the starting point of my scientific career. But the fantastic dreams of my youth, which inclined me to philosophy, seem to have maintained their persistent existence in the subconscious depths of my mind during my entire scientific life.

In describing my scientific career, six periods are to be distinguished: period I, 1929–1938; period II, 1939–1949; period III, 1950–1959; period IV, 1960–1967; period V, 1968–1973; period VI, 1973–1981. In each period there have been substantial changes in at least one of the following items: A, area of study and/or research; B, aim of research activities; C, affiliation; D, research group; E, main work and publications. In what follows I shall describe my scientific career in terms of A_x, B_x, C_x, D_x, and E_x where x runs through the six periods just mentioned.

2. Training To Be a Mathematician

This constituted my period I (1929–1938).

A_{I}: I started from mathematical analysis and made the transition to probability theory.

B_{I}: My scientific interests changed from an analysis of the cosmos to trying to understand chaos.

C_{I}: (i) I was first a student in the Department of Mathematics, Tokyo University, (1929–1934), then (ii) an assistant in the Department of Mathematics, Osaka University (September 1934–November 1938), and finally (iii) a lecturer in the Department of Exact Engineering in Osaka University (November 1938–March 1939).

D_{I}: In my period at Osaka University, I worked in close contact with many young analysts in Japan. I had the occasion to meet Professor Norbert Wiener on his way to China in 1935, and enjoyed the opportunity of consulting him on my thesis and having a talk on random walks with him.

E_{I}: My work was concentrated on functional equations of several types in [1], [2] (my doctoral thesis), and [3] under the influence of Professors M. Nagumo (functional equations) and S. Izumi (Fourier analysis).

After completing my work on functional equations, which amounted to 18 papers, I found myself unhappy about the direction of my research. I had become more and more inclined to see mathematics as a model of reality, and felt that deeper contact with reality was required to attain a truer model of it. In contrast with the ordered cosmos provided by mathematical analysis, there was a world of chaos whose mathematical modelling could at least partially be described by probability theory. With two young colleagues in Osaka University, K. Yoshida and S. Kakutani, both now outstanding mathematicians of our time, I joined a study group on the modern theory of probability, which had not been taught in any systematic way when we were mathematics students at universities in Japan.

3. Transition from Mathematics to Statistics

This formed the substance of my period II (1939–1949).

A_{II}: Starting from probability theory, I made a gradual transition to statistics.

B_{II}: I moved from the world of ideas to the world of reality.

C_{II}: I was appointed Assistant Professor at Kyushu University, in April 1939, with the responsibility of establishing with my two colleagues Professors M. Hukuhara and H. Honbu a new Department of Mathematics in the newly founded Faculty of Science.

I was particularly responsible for promoting applied differential equations and mathematical statistics. In spite of the high achievements of Japanese pure mathematics at that time, applied mathematics in general and mathematical statistics in particular had not been appropriately treated in any existing Department of Mathematics in Japan. My feeling was that Japan was then thirty years behind English-speaking countries such as the UK, the USA, and India, in the area of mathematical statistics.

In order to make up this thirty-year lag, I determined to become a self-taught student, without a teacher in statistics, simply assuming myself to be twenty-one again. Furthermore, I considered that it was indispensable to organize some research association to promote studies, surveys, and investigations on modern statistical theory and practice which had been so greatly developed in Europe and the United States. With this idea, the Research Association of Statistical Science (Tokei Kagaku Kenkyu Kai) was established in February 1941 with about 100 ordinary members coming from mathematics, statistics, economics, biology, actuarial science, medicine and engineering industries all over Japan. In 1942 the Association started to publish the *Bulletin of Mathematical Statistics* in English: this has continued to be published regularly since then.

D_{II}: In 1943, I became ordinary Professor of Mathematical Statistics in the Department of Mathematics, Kyushu University. I was a temporary Member of the Statistics Committee of the Japanese Government. From 1948 to 1949 I was an acting Director of the Institute of Statistical Mathematics in Tokyo.

E_{II}: A survey paper [4] on the theory of mutually independent stochastic variables and several works on the theory and application of weakly contagious stochastic processes [5], [6], [61], [62] constituted my start in statistics.

During this period I became interested in the application of quality control to industry. As early as 1941, I realized that one of the real bases for promoting modern statistics lay in the area of statistical quality control in industry. With the cooperation of Y. Ishida, who himself had a lot of experience of statistical quality control in his job of manufacturing electric lamps, I produced a Japanese translation of E. S. Pearson's monograph [64]. After World War II, I was very busy introducing statistical quality-control methods to engineers working in industry.

I was also concerned with designs of sample surveys. For example, I was involved in designing the sample surveys in [63], [65], which covered (a) cost-of-living surveys of coal-miners, (b) fishery catch volumes, and (c) timber volume. These experiences led me to introduce several statistical techniques including the design of experiments and time series analysis into these areas.

I was instrumental in the publication of statistical tables [66], [67]. Finally my monograph [7] *Recognition of Statistics* was published to make clear the historical background of statistics and the development of methodologies during the past 300 years. Such a work was indispensable for the Japanese community to familiarize itself with modern statistics.

4. Establishment of Statistical Theories; Domestic and International Contacts

These occupied me during my period III (1950–1959).

A_{III}: I tried to establish statistical theories through learning by accumulating experiences of the real world.

B_{III}: I attempted to develop international scientific activities and contacts by visiting statistical institutions abroad. I also took part in domestic scientific activities as a member of the Science Council of Japan from 1951 on.

C_{III}: In this period, I visited the Indian Statistical Institute as a visiting professor at the invitation of Professor P. C. Mahalanobis for four months from April to August 1953. Later, I returned to India as a member of the Reviewing Committee on the National Sample Survey in India for two months from December 1957 to January 1958. I was a visiting professor at Iowa State College at the invitation of Professor T. A. Bancroft for one semester from September to December 1955, and a Research Associate at the Statistical Laboratory, Princeton University where I was invited by Professor S. Wilks under a Rockefeller Foundation Grant from December 1957 to April 1958. It was also during this period that I made several short visits to Stanford University and the University of California at Berkeley.

D_{III}: I was fortunate at this time to make the acquaintance of many eminent statisticians throughout the world. It was on the occasion of the General Session of the International Statistical Institute (ISI) held in Rio de Janeiro, Brazil in July 1955, that I had my first stimulating talk with R. A. Fisher. He immediately took the initiative of recommending me for membership of the ISI. Following this initial meeting, I was able to meet and talk to him on several occasions in various parts of the world. P. C. Mahalanobis invited me to India for a second time in 1957.

With these two great statisticians, I had plenty of opportunity of hearing directly from them about their respective statistical ideas on the design of experiments, sample survey techniques and other topics. In fact, my papers [17], [18] are closely connected with what Mahalanobis called historical design and mapping problems, and were presented to Abraham Wald before his tragic death in South India in 1951. With regard to my idea on two-sample theory, R. A. Fisher wrote me a letter suggesting that it was similar to a viewpoint he had advocated in an old paper of his. After several talks with Fisher, I later published a paper [19] interpreting his statistical ideas.

It was also during this period that I learned cybernetics from Professor Norbert Wiener in Japan and Massachusetts, and this had a great influence on my subsequent scientific activities. Professor Jerzy Neyman was kind enough to invite me several times to the Berkeley Symposium and this led to the publication of my papers [23], [27], [28] and [29]. Professor Maurice Bartlett and Professor John Tukey were most capable Vice-Presidents when I was President of the International Statistical Association for the Physical Sciences, attached to ISI. I also enjoyed the friendship of Professor W. G. Cochran at Johns Hopkins University and Professor T. A. Bancroft at Iowa State University.

E_{III}: I had at that time four subjects of keen interest on which I was able to publish a series of papers. These were:

(1) Statistical inference processes: [9], [10], [11], [12], [14], [15] and five subsequent papers.

(2) Statistical control processes [13], [20], [21] and allied papers.
(3) Sample survey theory [17], [18].
(4) Design of experiments [16], [67], [68].

In these papers I was endeavouring to summarize my own experiences in statistical surveys and design of experiments through a theoretical formulation which was somewhat different from the more orthodox theories.

In addition to this research, I was busy publishing books and dictionaries in Japanese in order to increase the understanding of statistics in Japanese professional societies, both academic and business. These were:

(i) *Statistical Inference* I and II (Iwanami, 1958),
(ii) *Dictionary of Statistics*, of which I was editor, (Toyo Keizai Shimposha, 1951),
(iii) *Systematic Dictionary of Statistics*, of which I was also editor, (Toyo Keizai Shimposha, 1957),
(iv) *General Course of Statistics*, which I wrote with Michio Inaba, (Kyoritsu Shuppan, 1960).

This last book has had extensive circulation in Japanese universities since its first publication.

5. Planning of Scientific Research and Transition from Statistics to Information Science

During my next period IV (1960–1967), I went through the following stages.

A_{IV}: I developed summary views on the logical foundations of statistics and planned work leading to information science.

B_{IV}: I decided to devote myself to the establishment of long-range planning of scientific research in Japan.

C_{IV} and D_{IV}: (a) During the period from 1951 to 1968, except for 1960, I was a member of the Science Council of Japan. I was associated with the Committee on Long Range Research Planning during 1954–1968, and was chairman of the subcommittee on Information Science, during 1963–1968, as well as member of the Research Liaison Group in Statistics.

(b) For two terms from December 1961 to December 1967, I was Director of the Kyushu University Library.

(c) Within Kyushu University, I was a member of the Committee on Information Science whose Chairman was the President of the University.

(d) In 1961 I became a visiting professor at the University of Western Australia in Perth, by an arrangement with Dr. J. Gani who was then at the Australian National University, where I also spent some time. Since then I have been well acquainted with many statisticians in Australia, and particularly Canberra, including Professor P. A. P. Moran.

E_{IV}: My research work began to tend towards information science. My scientific work on statistics had been connected with control and design, whose features were not far removed from planning and programming. During Japan's struggle for the rehabilitation of scientific research activities after World War II and their further promotion, my proposal was to develop a long-range plan for scientific research covering all areas of the sciences, natural, social and human. My activities as a member of the Science Council of Japan are summarized in my monograph [32].

During this period, I was also responsible for promoting mathematical programming and computational mathematics in addition to mathematical statistics as I took up the first professorship of mathematical programming in the Department of Mathematics at Kyushu University. My researches were naturally concerned with the dynamic programming of Professor Richard Bellman, one of my most intimate friends. With Professors K. Kunisawa and S. Moriguchi, I organized a series of sessions on mathematical programming which continued for about ten years from1965 under the sponsorship of the Japanese Union of Scientists and Engineers.

Regarding statistical works, I had the opportunity to present an invited paper to the General Session of the International Statistical Institute held in Tokyo in 1960. This paper [24] aimed to summarize the background and fundamental principles under which my work on various topics had been developed during the 1950s. I was conscious of the fact that my research activities had almost attained a stable state. It so happened, however, that this consciousness of stability disappeared after three years, as can be observed from my three papers, one on the relativistic logic of mutual specification [25] and two on data analysis [26], [28].

To myself, my transition of interests from statistics to informatics appeared entirely logical and natural, because I had long been insisting upon the process of learning by experience, for which the notion of information is definitely indispensable. In 1967 I published in the Fifth Berkeley Symposium a paper [29] in this connection. Besides the mainstream of thought leading to a transition from statistics to information science, this period IV saw the continued publication of several papers in successive processes of statistical inferences and control following my ideas in period III. Among others, papers [22], [23] and [27] should be cited here in this connection.

6. Working in Information Science

Period V (1968–1973) was devoted essentially to developing my views on information science.

A_V: I worked on information science and social information science.

B_V: I tried to lay the logical foundations for an information science approach.

C_V and D_V: This period was my last stage as professor at Kyushu University from which, according to the retirement rules, I retired on 1 April 1973. Among my many positions at Kyushu University were (a) the Deanship of the Faculty of Science from July 1968 to August 1969 (b) the Directorship of the Research Institute of Fundamental Information Science during the period from January 1968 to March 1973. I also had several editorial responsibilities such as: (i) the editorship of the Information Science Series (in Japanese), published in 65 volumes by Kyoritsu Shuppan, (which began in 1968 and is still continuing) (ii) the editorship of the Social Information Science Series (in Japanese), published in 18 volumes by Gakushu Kenkyu Sha, (completed in 1979) (iii) the associate editorship of *Biomathematical Science* since 1966, and (iv) the associate editorship of *Information Science* from 1969.

E_V: My research work in this period can be classified into the following topics.

(a) *The first consists of the logical foundations for the information science approach by introducing a space of informative logics.* A representative paper [37] on this topic came from my lecture given at the International Symposium on Large Systems in Dubrovnik, Yugoslavia in August 1968. An extensive development of the ideas presented in this lecture was given in my monograph [30]. So far as the structure subspace is concerned, there is some resemblance between my logical formulations of statistics in [24] and the subspace of informative logics. I introduced the notion of *eizon* as a new key word in Japanese; "ei" is derived from the word "kei-ei" corresponding to management, while "zon" comes from "jitsuzon" implying existentialism. It should also be noted that one of our subspaces S_{Fe} has three coordinate systems (III_1) control, (III_2) eizon, and (III_3) creation.

(b) Our systematic approach to informative logics brought the topics of *creative engineering* and the logic of design into serious consideration as shown in [30], [35] and [42].

(c) Methodological considerations in *biomathematics* and their realization in terms of cell space approaches in biomathematics and neural dynamics were the subjects of my intensive studies during this period, as shown in my papers [36], [38], [43], [44], [46] and [50]. Fifteen papers associated with [43] were presented on these topics; some were written with M. Yamaguchi.

(d) The notion of "eizon" was applied in discussing the methodologies of environmental studies, as shown in our monograph [39] and the papers [40] and [47]. In combination with the biomathematical approaches enunciated in (c), I introduced the notion of *eizon ecosphere*, which turned out to be one of the key notions in my subsequent period VI.

(e) My final research topic in period V resulted in the publication of some expository monographs [32], [33], [34].

7. Scientific Activities in the International Institute for Advanced Study of Social Information Science (IIAS-SIS)

My period VI (1973–1981) has been spent at the IISA-SIS. Immediately after my retirement from Kyushu University, I was appointed to the directorship of the Institute which was established and sponsored by Fujitsu Limited, the Japanese electronics and communications company. As early as June 1972, I had the opportunity of presenting a memorandum to the President of Fujitsu, Mr. Kora, regarding the principles and policies under which such an institute should be established and managed. Following his complete agreement on my proposals, I was able to accept the responsibility of being the Director of IIAS-SIS. Since then, almost eight years have passed which represent period VI of my scientific career.

A_{VI}: I became interested in the formation of scientific information systems and the informatic analysis of research activities.

B_{VI}: On the basis of the informative logics developed during period V, I endeavoured to develop a systematic approach to the formation of scientific information systems [31], [45], [51], [52]. In this context I introduced the notions of *brainwares* [41], [45], [48] and *research automation* [51], [52], [54], [60].

C_{VI}: There were two successive Special Research Projects sponsored by the Japanese Ministry of Education, Science, and Culture: (i) Advanced Information Processing of a Large Amount of Data over a Broad Area, from April 1973 to March 1976; (ii) the Formation of Information Systems and the Organization of Scientific Information. In both these two Special Research Projects, I belonged to the Organizing Committees as well as being a research worker representing one Research Unit. My status in this connection was that of Professor Emeritus of Kyushu University, where I had worked for 34 years.

IIAS-SIS has been growing slowly but steadily; there are now twenty research workers working on ten research projects. These are (I) knowledge-based systems, (II) pattern recognition, (III) brainware systems, (IV) software engineering, (V) policy science, (VI) space development, (VII) system control, (VIII) non-linear programming, (IX) Josephson device cellular automation, and (X) machine translation of language.

D_{VI}: The first Special Research Project involved 200 scientists from representative universities all over Japan, while the second involved more than 400. With these scientists, I enjoyed very stimulating discussions about the topics of our concern. During this period I had numerous opportunities to visit many institutions working on computer science and informatics, and attended several international symposia and conferences in these areas. During the two years from June 1975 to May 1977, I accepted the responsibility of becoming President of the Information Processing Society of Japan for one two-year term.

E_{VI}: I have continued to work on the logical foundations of informatics, and endeavoured to generalize my ideas to the area of social information science in a series of my papers and monographs [49], [51], [52], [53], [55], [57], [58]. My participation in the two Special Research Projects led me to an informatic analysis of scientific research activities. As a result of this analysis, I have formulated the four basic research axes for promoting informatics; these consist of B-(I) the information processing process; B-(II) the space of informative logics; B-(III) the information expression scheme; B-(IV) intelligence levels; B-(V) the genesis and equilibrium among intelligence levels. In this connection, I pointed out the need for developing datalogy and semiology. With the rapid growth of knowledge engineering now in progress, I can but recall our experiences in the area of quality control. In connection with extensional and intensional data bases, I have introduced the notion of brainwares which are closely connected with the roles of testing hypotheses. Thus, I found myself returning to the area of statistical inference and control processes which, I find, can be generalized to respond to the needs of informatics.

Since 1977 I have been endeavouring to develop a mathematical formulation of the generalized relational ecosphere in a series of papers [56], [58], [59], which summarize my ideas on information-based societies.

8. Retrospect and Prospect

My scientific career can be broadly described as a sequence of transitions from mathematics in period I, to statistics in periods II, III, IV, and finally to informatics in the latest two periods V and VI, as I have indicated in the previous sections. In spite of these surface phenomena, my scientific itinerary has seemed to me to be a continuous pursuit of unsolved problems which presented themselves to me as I proceeded in life. I am now conscious that my local space is surrounded not only by mathematics, statistics, and informatics but also by cybernetics, logics, and linguistics. My present concerns are with semiology which refers back to my concern with the cosmos, and with datalogy which deals with chaos; from these I have learned much, and I hope that my experiences in the six periods of my career will contribute something to science. This retrospect and prospect reminds me of my youthful dreams, and show how much I have owed to my teachers, to my colleagues, and to my students, of whom I have mentioned only a few.

Although my scientific career has apparently been a sequence of somewhat drastic transitions, I have been blessed with a stable and happy family life. In this connection my sincere thanks are due to my wife, Misao, for her constant support. She came from Sendai as the daughter of the late

Professor Shirota Kusakabe, one of the founders of physical seismology and astronomy in Japan. We have three daughters, Kiyoko, Yoshiko, Shigeko and one son, Genshiro. All of them have married scientific research workers. My second daughter Yoshiko Takenaka has become a mathematician working in graph theory. My son Genshiro Kitagawa is working in time series analysis as a statistician. My wife and I now look forward to the future of our eight grandchildren, some of whom we hope may become scientists working in the area of mathematics, statistics, and informatics. Needless to say, I have every confidence that many students of mine will be successful in guiding groups of excellent scholars to broader areas of science including mathematics, statistics, and informatics.

Besides many of my students working in universities and industries, located in the central parts of Japan such as the Tokyo and Osaka areas, there is a remarkable concentration of them in Kyushu University where I have been Professor Emeritus since 1973. Perhaps I can mention Professors A. Kudo (mathematical statistics), N. Furukawa (mathematical programming) and T. Yanagawa (mathematical statistics) in the Department of Mathematics; Professors S. Kano (fundamental informatics), C. Asano (computational statistics) and S. Arikawa (computational informatics) in the Research Institute of Fundamental Information Science; Professors T. Kitahara (management science), M. Kodama (scheduling theory), and S. Iwamota (dynamical programming) in the Faculty of Economics; Professors K. Kono (experimental design), K. Sakaguchi (mathematical statistics) and T. Hamachi (probability theory).

In conclusion, I should mention that a series of books is now in preparation summarizing my research work during the four periods from III to VI mentioned earlier.

Publications and References

[1] Kitagawa, T. (1936) On the associative functional equations and a characterization of general linear spaces by the mixture-rule. *Japanese J. Math.* **13**, 435–458.

[2] Kitagawa, T. (1937) On the theory of linear translatable functional equations and Cauchy's series. *Japanese J. Math.* **13**, 233–332.

[3] Kitagawa, T. (1940) The characterizations of the fundamental linear operations by means of the operational equations. *Mem. Fac. Sci. Kyushu Univ.* A **1**, 1–28.

[4] Kitagawa, T. (1940) Theory of mutually independent stochastic variables (in Japanese). *Proc. Physico-Math. Soc. Japan* **14**, 236–254; 255–295.

[5] Kitagawa, T. (1940) The limit theorems of the stochastic contagious processes. *Mem. Fac. Sci. Kyushu Univ.* A **1**, 167–194.

[6] Kitagawa, T. (1941) The weakly contagious discrete stochastic process. *Mem. Fac. Sci. Kyushu Univ.* A **2**, 37–65.

[7] Kitagawa, T. (1948) *The Recognition of Statistics Foundation and Methodologies* (in Japanese). Hakuyosha. (New edition 1968.)

[8] KITAGAWA, T. (1950) *Tables of Poisson Distribution with Supplementary Note on Statistical Analysis of Rare Events* (in Japanese). Baifukan. (2nd edition 1968; English edition 1952; Dover edition 1953).

[9] KITAGAWA, T. (1950) Successive process of statistical inference (1). *Mem. Fac. Sci. Kyushu Univ.* A **5**, 139–180.

[10] KITAGAWA, T. (1951) Successive process of statistical inferences (2). *Mem. Fac. Sci. Kyushu Univ.* A **6**, 54–95.

[11] KITAGAWA, T. (1951) Successive process of statistical inferences (3). *Mem. Fac. Sci. Kyushu Univ.* A **6**, 131–155.

[12] KITAGAWA, T. (1951) Successive process of statistical inferences (4). *Bull. Math. Statist.* **5**, 35–50.

[13] KITAGAWA, T. (1952) Successive process of statistical controls (1). *Mem. Fac. Sci. Kyushu Univ.* A **7**, 13–26.

[14] KITAGAWA, T. (1953) Successive process of statistical inferences (5). *Mem. Fac. Sci. Kyushu Univ.* A **7**, 81–106.

[15] KITAGAWA, T. (1953) Successive process of statistical inferences (6). *Mem. Fac. Sci. Kyushu Univ.* A **8**, 1–29.

[16] KITAGAWA, T. (1956) *Lectures on Design of Experiments* (in Japanese), I, II. Baifukan.

[17] KITAGAWA, T. (1954) Some contributions to the design of sample surveys. *Sankhyā* **14**, 317–362.

[18] KITAGAWA, T. (1956) Some contributions to the design of sample surveys. *Sankhyā* **17**, 1–36.

[19] KITAGAWA, T. (1957) Successive process of statistical inference associated with an additive family of sufficient statistics. *Bull. Math. Statist.* **7**, 92–112.

[20] KITAGAWA, T. (1959) Successive process of statistical controls (2). *Mem. Fac. Sci. Kyushu Univ.* A **13**, 1–16.

[21] KITAGAWA, T. (1960) Successive process of statistical controls (3). *Mem. Fac. Sci. Kyushu Univ.* A **14**, 1–33.

[22] KITAGAWA, T. (1961) A mathematical formulation of the evolutionary operation program. *Mem. Fac. Sci. Kyushu Univ.* A **15**, 21–71.

[23] KITAGAWA, T. (1961) Successive process of optimizing procedures. *Proc. 4th Berkeley Symp. Math. Statist. Prob.* **1**, 407–434.

[24] KITAGAWA, T. (1961) The logical aspect of successive processes of statistical inferences and controls. *Bull. Internat. Statist. Inst.* **38**, 151–164.

[25] KITAGAWA, T. (1963) The relativistic logic of mutual specification in statistics. *Mem. Fac. Sci. Kyushu Univ.* A **17**, 76–105.

[26] KITAGAWA, T. (1963) Automatically controlled sequence of statistical procedures in data analysis. *Mem. Fac. Sci. Kyushu Univ.* A **17**, 106–129.

[27] KITAGAWA, T. (1963) Estimation after preliminary test of significance. *Univ. Calif. Publ. Statist.* **3**, 147–186.

[28] KITAGAWA, T. (1965) Automatically controlled sequence of statistical procedures. *Bernoulli, Bayes, Laplace Anniversary Volume*, Springer-Verlag, Berlin, 146–178.

[29] KITAGAWA, T. (1968) Information science and its connection with statistics. *Proc. 5th Berkeley Symp. Math. Statist. Prob.* **1**, 491–530.

[30] KITAGAWA, T. (1969) *The Logic of Information Science* (in Japanese). Kodansha.

[31] KITAGAWA, T. (1969) Information science approaches to scientific information systems and their implication to scientific researches. *Res. Inst. Fund. Informat. Sci. Kyushu Univ. Res. Rep.* **3**, 1–41.

[32] KITAGAWA, T. (1970) *Way to Science Planning—Seventeen Years in the Science Council of Japan* (in Japanese). Kyoritsu Shuppan.

[33] KITAGAWA, T. (1970) *Thirty Years in Statistical Science—My Teachers and My Friends* (in Japanese). Kyoritsu Shuppan.

[34] KITAGAWA, T. (1970) *Viewpoint of Information Science—Search for a New Image of the Sciences* (in Japanese). Kyoritsu Shuppan.

[35] KITAGAWA, T. (1970) The logic of design (in Japanese). In *Design Series in Architecture* **4**, *Creation in Design*. Hudosha.

[36] KITAGAWA, T. (1971) A contribution to the methodology of biomathematics —information science approach to biomathematics. *Math. Biosci.* **12**, 25–41.

[37] KITAGAWA, T. (1972) Three coordinate systems for information science approaches. *Information Sci.* **5**, 157–169.

[38] KITAGAWA, T. (1973) Dynamical systems and operators associated with a single neuronic equation. *Math. Biosci.* **18**, 191–244.

[39] KITAGAWA, T. (1973) Environments and eizonsphere (in Japanese). *Social Information Science Series* **16** (1), Gakushu Kenkyu Sha, 132–199.

[40] KITAGAWA, T. (1974) The logic of information sciences and its implication for control process in large system. *Second FORMATOR Symp. Math. Methods for Analysis of Large Scale Systems, Prague*, 13–31.

[41] KITAGAWA, T. (1974) Brainware concept in intelligent and integrated system of information. *Res. Inst. Fund. Informat. Sci. Kyushu Univ. Res. Rep.* **39**, 1–20.

[42] KITAGAWA, T. (1974) *Historical Image of Human Civilization—a Prolegomena to Informative Conception of History* (in Japanese). Social Information Science Series **17** (1), Gakushu Kenkyu Sha.

[43] KITAGAWA, T. (1974) Cell space approaches in biomathematics. *Math. Biosci.* **19**, 27–71.

[44] KITAGAWA, T. (1975) Dynamical behaviours associated with a system of *N* neuronic equations with time lag. *Res. Inst. Fund. Informat. Sci. Kyushu Univ. Res. Rep.* **46**, 1–59; **49**, 1–87; **54**, 1–97.

[45] KITAGAWA, T. (1975) Statistics and brainware in intelligent and integrated information system. *Res. Inst. Fund. Informat. Sci. Kyushu Univ. Res. Rep.* **58**, 1–13 (*Proc. 40th Session Internat. Statist. Inst.* **4**, 30–40).

[46] KITAGAWA, T. (1975) Dynamics of reverberation cycles and its implications to linguistics. *Res. Inst. Fund. Informat. Sci. Kyushu Univ. Res. Rep.* **56**, 1–58.

[47] KITAGAWA, T. (1976) Some contributions to the methodology of system scientific approach to environmentology. *Res. Inst. Fund. Informat. Sci. Kyushu Univ. Res. Rep.* **71**, 267–284.

[48] KITAGAWA, T. (1977) Structure formation of scientific brainware as a basis of data management. *Bull. Math. Statist.* **17**, 54–64.

[49] KITAGAWA, T. (1977) *World Views of Information Science* (in Japanese). Diamond Co.

[50] KITAGAWA, T. (1978) Neural equation approaches to recognition processes on the basis of reverberation and virtual dynamics. *Internat. Symp. Math. Topics in Biology*, 121–130.

[51] KITAGAWA, T. (1978) Formation of scientific information system and its implication to informatics. *J. Information Processing* **1**, 111–124.

[52] KITAGAWA, T. (1978) A theoretical formulation of information network system and its implications to social intelligence. *Internat. Inst. Adv. Study Social Information Sci. Res. Rep.* **1**, 1–21.

[53] KITAGAWA, T. (1978) The roles of computerized information systems in knowledge societies. Keynote address. *8th Australian Computer Conference, Canberra* III, XX–XLII.

[54] KITAGAWA, T. (1978) Informatical analysis of scientific research. *Res. Inst. Func. Information Sci. Kyushu Univ. Res. Rep.* **81**, 1–52.

[55] KITAGAWA, T. (1978) Mathematical approaches in theoretical cybernetics

and informatics. *8th Internat. Congr. Cybernetics, Namur*, 267–284.

[56] KITAGAWA, T. (1978) An informatical formulation of generalized relational ecosphere on the basis of paired categories. *Internat. Conf. Cybernetics and Systems*, 322–327.

[57] KITAGAWA, T. (1979) Informational approach to models and architectures for distributed data base sharing systems. *Seminar on Distributed Data Sharing* Systems. Inst. Nat. de Recherche en Informatique et en Automatique,

[58] KITAGAWA, T. (1979) The logics of social information science. *Social Information Science Series* **18** (1), Gakushu Kenkyu Sha, 26–137.

[59] KITAGAWA, T. (1979) Generalized artificial grammars and their implications to knowledge engineering approaches. *Internat. Inst. Adv. Study Social Informat. Sci. Res. Rep.* **6**, 1–29.

[60] KITAGAWA, T. (1980) Some methodological consideration on research automation. (I) objectivization and operatorization. *Res. Inst. Fund. Informat. Sci. Res. Rep.* **96**, 1–31.

[61] KITAGAWA, T. and HURUYA, S. (1940) The application of limit theorem of the contagious stochastic process to the contagious diseases. *Mem. Fac. Sci. Kyushu Univ.* A **1**, 195–207.

[62] KITAGAWA, T., HURUYA, S. and YAZIMA, T. (1941) The probabilistic analysis of the time series of rare events, I. *Mem. Fac. Sci. Kyushu Univ.* A **2**, 151–204.

[63] KITAGAWA, T. and HUZITA, T. (1951) *Sample Survey on Living Costs – Living Conditions of Coal Mining Labourers* (in Japanese). Toyo Keizai Shimposha.

[64] KITAGAWA, T. and ISHIDA, Y. (1942) Japanese translation of *The Application of Statistical Methods to Industrial Standardization and Quality Control* by E. S. Pearson, British Standards Inst. Publ. 600 (1935).

[65] KITAGAWA, T., KINASHI, K. and NISHIZAWA, M. (1962) New development of sampling designs in forest inventories. *Bull. Kyushu Univ. Forests* **35**, 1–82.

[66] KITAGAWA, T. and MASUYAMA, M. (1942) *Statistical Tables*, I. Research Association of Statistical Science, Kawade Shobo.

[67] KITAGAWA, T. and MITOME, M. (1953) *Tables for the Design of Factorial Experiments* (in Japanese). Baifukan.

[68] KITAGAWA, T. and MITOME, M. (1963) Group theoretical tabulations of confounded factorial designs with reference to "Tables for Designs of Factorial Experiments." *Misc. Bull. Kyushu Agricult. Exper. Statist.* **3**, 1–107.

5
STATISTICIANS IN INDUSTRY AND ECONOMETRICS

L. H. C. Tippett

Leonard Henry Caleb Tippett was born in London in 1902. He completed his secondary education at the St. Austell County School, Cornwall, in 1920, and went on to read physics at Imperial College, London, from which he graduated in 1923. For the following two years, he studied statistics with Karl Pearson at University College, London and then took up a position as statistician at the British Cotton Industry Research Association (Shirley Institute) in 1925. Before he retired in 1965 he had risen to become one of its assistant directors.

Leonard Tippett has taken an active role in statistical developments in the UK, and particularly the application of statistics to the textile industry. He was elected President of the Manchester Statistical Society in 1960, and later of the Royal Statistical Society (RSS) in 1965. Among his many honours are the Guy Silver Medal of the RSS, and the Shewhart Medal of the American Society for Quality Control. He is the author of three books and fifteen papers on statistics.

He was married in 1931, and has three sons. His hobbies are music, gardening and walking.

The Making of an Industrial Statistician

L. H. C. Tippett

1. My Early Education

The foundation of my education was laid in the public elementary school (as it was then named) at Mount Charles, St. Austell, Cornwall, where we were taught the three Rs by old-fashioned methods. I remember that when teaching the arithmetic of stocks and shares the irascible teacher designated his left-hand trousers pocket for capital and his right-hand one for income. He would detail some transaction and then would ask a pupil into which pocket the proceeds should be put. The less competent pupil would make a frantic guess, and if he got it wrong there would be a rap on the desk from a cane, and a shout of "you blockhead." Under this rough pedagogy I attained a good mastery of arithmetic, and in 1913 at the age of 11 won a scholarship to St. Austell County School.

Secondary education had not developed, particularly in remoter parts of the country such as Cornwall, as it has done since, and education was also affected by war-time shortages and staffing difficulties, for most of my secondary schooling was during the First World War. The pupils all followed much the same curriculum, but when towards the end a university course seemed to be possible, a specialism had to be chosen. The only science options available were chemistry and physics, and as I felt that my reasoning powers were stronger than my memory, I chose the latter. Mathematics to honours standard was beyond our reach.

2. Student Days in London

In 1920 I became a student in the physics department of the Royal College of Science, London (a constituent of Imperial College and a school of London University). The head of department was Professor H. L. Callendar of steam tables fame. He was rather remote from us undergraduates; we suspected that he avoided us by retiring into an inner sanctum when he heard a knock on his study door—he had no secretary to organise his appointments. But doubtless mucn of the character of the curriculum was due to him. We studied largely the classical physics of heat, light, sound and electricity. We were given some introduction to the newer atomic physics of the early 1920s, but were not encouraged to go whoring after that harlot: during the whole course I never handled a thermionic valve in the laboratory. The practical bent of our training was enhanced by short courses in elementary mechanical and electrical engineering. The whole was an excellent training for one who was going to work in industry, although at the time I did not know that that was to be my destiny.

Another feature of the course, the good influence of which persisted, was an insistence on the accurate measurement of physical quantities, and a disciplined approach to the scientific controversies of the day. Perhaps less valid was the way in which we tended to think of a hypothesis as being either right or wrong, and regarded the crucial experiment as being the main instrument for advancing knowledge. When later I did a sampling experiment to test some theoretical result and found that the experimental results did not agree with theory, I found it better to improve the sampling technique than to discard the theory. Life is too complex to be fully described by such simple theoretical models as we are able to construct. They may contain some truth, even if they are not completely correct.

After graduating in physics I was awarded a studentship by the (then) British Cotton Industry Research Association, known as the Shirley Institute, to study statistics under Karl Pearson at University College, London, and later for a short time under R. A. Fisher (later Sir Ronald Fisher) at Rothamsted. The research workers at the Shirley Institute were finding much uncontrolled variation in the fibre, yarn and fabric properties they were measuring and it was thought that statistical methods, which were only beginning to have currency, would be helpful. I suspect that I was chosen partly because of an undeserved reputation for a competence in mathematics. Any eminence I had in this connection was due to the flatness of the surrounding country; I have always regretted my weakness in mathematics.

The time at University College was thrilling. Karl Pearson was a great man, and we felt his greatness. He was hard-working and enthusiastic and inspired his staff and students. When I was there he was still doing research, and would come into his lectures, full of excitement and enthusiasm, giving results hot from his study table. That in those years his lines of research were somewhat out of date did not make his lectures any less stimulating. As a young man he had followed many interests, one notable product being *The Grammar of Science*. It was typical of his breadth of interests that one of his lecture courses was on "The History of Statistics in the 17th and 18th Centuries" (these have recently been published by his son, the late Egon S. Pearson). The department was developing the relatively new sciences of eugenics and biometrics, in connection with which he had had close personal contact with Sir Francis Galton. As a result, the embers of controversies over the laws of inheritance were still glowing. K. P. still favoured regression analysis for studying the relations between generations and had little time for Mendelism. He was a vigorous controversialist. An example giving the flavour of his style is in the preface to the first edition of *Tables for Statisticians and Biometricians* where he wrote "It is a singular phase of modern science that it steals with a plagiaristic right hand while it stabs with a critical left"; and one series of publications issued by him is termed "Questions of the Day and of the Fray."

When I was at University College the influence of the vigorous and controversial past was in the atmosphere. The walls of the department were embellished with mottoes and cartoons. One motto was Lord Kelvin's frequently quoted approbation of the use of measurement and numerical expression; there was also a cartoon by 'Spy' which was a caricature of 'Soapy Sam'—the Bishop Wilberforce who had the famous verbal duel with T. H. Huxley on Darwinism at the British Association meeting in 1860. There was a display of publications that had been issued over past decades, and an impression of the interests of the department was given by such titles as "Treasury of Human Inheritance (Pedigrees of Physical, Psychical, and Pathological Characters in Man)" and "Darwinism, Medical Progress and Eugenics." K. P. reminded us of his close connection with Galton at an annual departmental dinner when he gave a description of the year's work in the form of a report such as he would have given to Galton, had he been alive. And we toasted "the biometric dead."

For me this atmosphere was specially congenial because the work had a human element; for the impersonality of physics had left me emotionally unsatisfied, in spite of the interest of the subject. But most satisfying of all was to go from an exact science in which uncontrolled variation and its associated uncertainties were merely a nuisance, to the new concepts and methods of statistics. Like Keats "On First Looking into Chapman's Homer,"

> Then felt I like some watcher of the skies
> When a new planet swims into his ken;
> Or like stout Cortez when with eagle eyes
> He star'd at the Pacific—and all his men
> Look'd at each other with a wild surmise—
> Silent, upon a peak in Darien.

My contacts with Fisher came after I had become familiar with statistics, and were relatively brief. Nevertheless I was very taken with the ideas and methods he was developing and applying for dealing with small samples, and in my first book [7], published in 1931, I attempted to present a unified system of statistics, uniting the approaches of Pearson and Fisher.

3. My Work at the Shirley Institute

On leaving University College in 1925 I joined the staff of the Shirley Institute, a fully qualified but woefully inexperienced statistician. I had much to learn. Colleagues either thought that statistics was of no use to them, or thought of a statistician as a magician. One colleague showed me a table of fibre properties and asked me to treat them statistically—it was some time before I realised that one should start at the other end and consider first the technical implications of the data, using statistical methods to give form to the technical measures and to validate conclusions. I had adopted the statistical approach so completely that I regarded extreme individuals only as part of the natural variation. It was only later that I realised that extreme individuals could result from the operation of technical causes and that it might be worth studying those individuals specially so as to discover these causes.

In the 1920s it was known that looms in weaving sheds worked for only about 70 per cent of their time, but the causes of the lost 30 per cent were largely unknown. This was thought to be a suitable topic for a statistician to study, and I was assigned the task. At first I thought of a purely statistical approach, measuring various quantities for a few hundred mills and correlating them in the hope of getting a lead on the causes of variations in productivity. This led nowhere, and the investigation became what would later be recognised as production engineering studies of the operation of the looms, listing the causes and frequencies and durations of loom stoppages. This was not very exciting statistically, for simple tables and averages gave all the relevant information, but it was effective. One piece of information that emerged was that looms were sometimes stopped awaiting the attention of the weaver who was busy attending to another loom under her charge—a subject for the application of queuing theory. It was while doing these surveys that I happened upon the work-study technique now known as work sampling.

The Shirley Institute had use for the standard statistical methods applicable wherever there is uncontrolled variation—averages, inference from samples, analysis of variance, design of experiments, and so on. I acted as a statistical consultant to colleagues, but tended to put them into the way of doing their own statistics rather than doing their statistics for them. We never had a statistics department. This perhaps resulted in relatively unsophisticated analyses, but it was a gain that the physicists, chemists, and technologists fully understood what they were doing.

Some experiments conducted in the laboratories and mills gave scope for the application of standard statistical designs. We had the common experience that the design was conditioned by technical and other practical considerations. We did not go in for complicated factorial designs. This was partly because random errors were large so that "treatments" had to be very few, and also the design had to be one that the experimenter could administer and understand. It may be that we would have used more adventurous designs had I been more competent in handling them.

One special statistical problem concerned the relationship between the length of the test specimen of, say, a yarn and the strength. If the length of the yarn can be regarded as a sample of n elemental lengths, the strength of the whole length is that of the weakest element—i.e., the value of the weakest of a sample of n. This phenomenon had been studied in these terms by the late Dr. F. T. Peirce before I joined the Institute, and although he did not use the orthodox statistical terminology, it was interesting to see the parallel to work I had been doing at University College on the extreme individuals in samples of n. Incidentally, the model I have described is too simple to fit the practical situation completely, but it describes an important part of the effects observed in textile testing.

Another special problem was the study of the variation of thickness along the length of a yarn. For a staple yarn a plot of thickness against length in effect forms a time series: it is assumed that thickness is measured at equal intervals along the length. In this time series are superimposed: periodic variations arising from imperfections in the spinning machinery; variations that appear to be wave-like but do not have a constant wave-length, which are a fundamental result of the process; and random variations. These different forms of variation produce different effects in the resulting cloth. Thus the analysis of time series was an important subject to which members of the Institute staff made a contribution.

As time went on more and more members of the staff became familiar with statistical ideas and methods, or such as they needed in their work, and my interests were deflected into textile technology and some aspects of industrial management. Quality control and labour productivity were the two main subjects of interest. During the Second World War and in the early post-war years many sectors of British industry were regarded as deficient in these matters, and vigorous propaganda and educational campaigns were instituted. I participated in these. Naturally, my main activities were with the Lancashire textile industry, but I was also encouraged to help other industries, and for a time was seconded as a part-time consultant to the (then) Anglo-American Council on Productivity.

4. Concluding Remarks

I remained on the staff of the Shirley Institute until retirement, moving up the administrative ladder until towards the end I almost ceased to be a statistician. Perhaps I may claim the title of Statistician Emeritus.

Of my published statistical work the only items that are likely to survive are "On the extreme individuals and the range of samples taken from a normal population" [3] and *Random Sampling Numbers* [5]. These are both student efforts, done under the close guidance of Karl Pearson. Most of my other publications have been textbooks and expository papers applying standard statistical methods to industrial, especially textile, fields. These, I hope, have served their day, and are now largely obsolete.

Publications and References

[1] Fisher, R. A. and Tippett, L. H. C. (1928) Limiting forms of the frequency distribution of the largest or smallest member of a sample. *Proc. Camb. Phil. Soc.* **24**, 180–190.

[2] Pearson, K. and Tippett, L. H. C. (1924) On the stability of the cephalic indices within the race. *Biometrika* **16**, 118–138.

[3] Tippett, L. H. C. (1925) On the extreme individuals and the range of samples taken from a normal population. *Biometrika* **17**, 364–387.

[4] Tippett, L. H. C. (1926) On the effect of sunshine on wheat yield at Rothamsted. *J. Agric. Sci.* **16**, 159–165

[5] Tippett, L. H. C. (1927) *Random Sampling Numbers*. Tracts for Computers No. 15, Cambridge University Press, London.

[6] Tippett, L. H. C. (1930) Statistical methods in textile research. Part 1, The analysis of complex variations. *J. Textile Inst.* **21**, T105–126.

[7] Tippett, L. H. C. (1931) *Methods of Statistics*. Williams and Norgate, London. (Second edition 1952).

[8] Tippett, L. H. C. (1932) A modified method of counting particles. *Proc. R. Soc. London* A **137**, 434–446.

[9] Tippett, L. H. C. (1935) Some applications of statistical methods to the study of variation of quality in the production of cotton yarn. *J.R. Statist. Soc. Suppl.* **2**, 27–55.

[10] Tippett, L. H. C. (1935) Statistical methods in textile research. Part 2, Uses of the binomial and Poisson distirbutions. *J. Textile Inst.* 26, T13–50.

[11] Tippett, L. H. C. (1935) Statistical methods in textile research. Part 3. A snap-reading method of making time-studies of machines and operatives. *J. Textile Inst.* **26**, T51–70.

[12] Tippett, L. H. C. (1943) *Statistics*. Oxford University Press, London. (Second edition 1968).

[13] Tippett, L. H. C. (1944) The control of industrial processes subject to trends in quality. *Biometrika* **33**, 163–172.

[14] Tippett, L. H. C. (1944) The efficient use of gauges in quality control. *Engineer, London* **177**, 481.

[15] Tippett, L. H. C. (1947) The study of industrial efficiency, with special reference to the cotton industry. *J.R. Statist. Soc.* **110**, 108–122.

[16] Tippett, L. H. C. (1950) *Technological Applications of Statistics*. Wiley, New York.

[17] Tippett, L. H. C. and Vincent, P. D. (1953) Statistical investigations of labour productivity in cotton spinning (with discussion). *J.R. Statist. Soc.* A **116**, 256–272.

[18] Tippett, L. H. C. (1969) *A Portrait of the Lancashire Textile Industry*. Oxford University Press, London.

[19] Tippett, L. H. C. and Main, V. R. (1941) Statistical methods in textile research. Part 4. The design of weaving experiments. *J. Textile Inst.* **32**, T209.

Herman Wold

Professor Herman Wold was born on 25 December 1908 in Skien, Norway; he was the youngest of a family of six brothers and sisters. In 1912, his parents migrated to Sweden; it was there that he received his school education before becoming a student at the University of Stockholm. He obtained his doctoral degree from Stockholm in 1938, after studying under Professor Harald Cramér.

His research was initially concerned with the econometric analysis of consumer demand, and was later directed to path models with directly observed and latent variables. Throughout his life, Professor Wold's work has been inspired by real-world applications; he has published over 200 scientific papers and some ten books.

In 1942, he was appointed to the University of Uppsala as Professor of Statistics, and remained in this position until 1970 when he moved to the University of Gothenburg. He retired as Professor Emeritus in 1975.

Professor Wold's many contributions to statistics and econometrics have been recognized by election to Fellowship of the Institute of Mathematical Statistics in 1946, the American Statistical Association in 1951, Honorary Fellowship of the Royal Statistical Society in 1961, and an Honorary Doctorate from the Technical University of Lisbon in 1965. He was elected to membership of the Swedish Academy of Sciences in 1960, and served as a member of the Nobel Economic Science Prize Committee from 1968 to 1980.

Professor Wold has also been active in international statistical and econometric affairs. He served as Vice-President of the International Statistical Institute from 1957 to 1961, was elected to Presidency of the Econometric Society in 1966, and Honorary Membership of the American Economic Association and the American Academy of Arts and Sciences in 1978. He has visited many universities in Europe and North America; most recently in 1978–80 and again in 1981–82 he has been a guest professor at the University of Geneva.

Professor Wold is a man of wide interests, scientific, musical, artistic and literary. He has been an active sportsman all his life: tennis, golf, walking, jogging, and skiing have kept him fit. He is married and has one son and two daughters.

Models for Knowledge

Herman Wold

Dedicated to Harald Cramér

1. Introduction

Professor Gani's invitation to contribute to the present volume has given me a welcome opportunity to review my work and discuss the research process.

Section 2 of this paper reviews my scientific work as it evolved during four main periods. These consist of my undergraduate studies and my doctoral thesis on stationary time series, 1932–38; econometric analysis of consumer demand, 1938–52; econometrics and path models with directly observed variables, 1945–70; and finally systems analysis, and path models with indirectly observed variables, from 1966 onwards. Throughout these 50 years, my contributions have been based on the least squares (LS) principle.

Since the 1930s the mainstream of contemporary statistics has been the maximum likelihood (ML) principle. ML is a highly developed body of methods, all based on the assumption that the observables are ruled by a joint probability distribution, usually a multivariate normal distribution. In the triumphant ML evolution, LS was left in a backwater. LS being an autonomous optimization principle, it suddenly dawned upon me that the rationale of LS could be established on a distribution-free basis, by subjecting the predictive relations of the model to predictor specification (1959–63). In the ensuing LS framework I initiated first the fix-point (FP) method for estimation of interdependent systems (1965, 1966) and then the partial least squares (PLS) method of path models with latent variables indirectly observed by multiple indicators (1971–72). The PLS approach, or briefly soft modelling, has kept me busy ever since.

Colleagues and friends honoured my sixtieth birthday by a Festschrift entitled *Scientists at Work* [4]; in this they reported on their personal experiences of research as a creative process. Being highly interested in this important but intricate theme, I have over the years collected notes to supply it with a *post festum* chapter of my own. These form the substance of Section 3 of this paper.

Section 3 links up with earlier insights on the research process. Sir Peter Medawar's outstanding book *Advice to a Young Scientist* [8], where the research process is characterized as *imaginative guesswork*, is referred to; the characteristic feature of a fruitful research milieu is seen to be its *synergy*, the joint participation of research workers in discussion of their various projects. The term "Hoeffding effect" refers to the Danish psychologist [6]. There is some discussion of the motivation for research and the selection of research topics, distinguishing opportunity research from the gold-rush phenomenon and from self-initiated research.

Section 4 comments on the present shortcomings of quantitative methods in systems analysis and systems evaluation. The research system as a pivotal part of our social structure is discussed in the light of the challenging economic–historical perspectives of Douglass North [9]. On both sides of the Atlantic the current research performance of the university system shows symptoms of bureaucratic inefficiency; this calls for determined efforts to improve matters by systems analysis and systems evaluation.

2. An Overview of My Scientific Work 1932–1981

2.1. 1932–38. Undergraduate Studies and My Doctoral Thesis (1938)

Swedish population statistics, unique in that they date back as far as 1750, show that mortality rates up to the 1830s fluctuated around the 30 per cent level; they then gradually decreased as welfare increased in the course of industrialization, to stabilize after 1900 at the 10 per cent level. In 1934 Harald Cramér and I analyzed the evolution of mortality, using the Makeham–Gompertz law to represent the dependence on age; separate analyses for males and females, and for mortality changes by 5-year calendar periods and 5-year birth cohorts, are reported. The Swedish association of local fire insurance companies asked me to propose a tariff system for their planned co-operative reinsurance company. In 1936, I designed the tariff rate to be proportional to the average local risk plus a safety supplement proportional to the standard deviation of the local risk rate. This design for the tariff is still being used; for details see Borch [3].

The main contribution in my doctoral thesis [10] is the "decomposition theorem,"

$$y_t = p_t + x_t, \qquad t = 0, \pm 1, \pm 2, \ldots \tag{1}$$

where the given stationary series y_t is decomposed into the "deterministic" component p_t which can be forecast exactly by its past $p_s (s < t)$, and the "stochastic" part x_t which is a moving sum of mutually uncorrelated "innovations" e_t such that

$$x_t = \sum_{s=0}^{\infty} (c_s e_{t-s}). \tag{2}$$

Forming a sequence of autoregressive series of the type

$$y_t = b_1 y_{t-1} + b_2 y_{t-2} + \cdots + b_h y_{t-h} + e_t^{(h)}, \tag{3}$$

the innovations e_t are obtained in the limit as $h \to \infty$.

Applications in my thesis included: (i) a model for the yearly level y_t of Lake Vener as a moving summation (2) of the rainfalls e_t and e_{t-1}, and (ii) an autoregression model (3) for business cycles y_t in Sweden from 1830 to 1913. In both models (i)–(ii) the deterministic component was absent, $p_t = 0$.

My main incentive for the decomposition (1)–(2) of a general stationary time series was to cover as special cases the three types of stationary time series that were known at the time. These were (i) the scheme of "hidden periodicities," where p_t is a sum of cosine functions with different periods, and x_t is a random disturbance; (ii) the moving summation model (2) with finite s; and (iii) the linear autoregression (3) with finite h. The periodicity of tidal water is an ancient application of (i). Moving summations (ii) of random numbers were used by G. U. Yule in 1921 and E. E. Slutsky in 1927 to generate quasi-cyclical variation. Distributed lags were introduced in regression analysis by Irving Fisher in 1925. Yule's famous 1927 model for the periodicity of sunspots, with its slightly irregular period of 11 years, is of the autoregressive type (iii).

When establishing the decomposition (1)–(2) I was only vaguely aware of its close affinity to the spectral analysis of time series, and even less aware of the ensuing affinity to operator theory. It so happened that the decomposition (1)–(2) gave a clue to overcoming an earlier impasse in spectral analysis and operator theory, as was shown by André Kolmogorov in 1939 and Norbert Wiener in 1942. Thanks to this profound aspect, my decomposition is referred to in the *Encyclopedia Britannica*, 13th edition, in the article on "Automata"; see also Helson [5].

In the summer before presenting my doctoral dissertation I was appointed by a government committee to carry out an econometric analysis of consumer demand on the basis of available Swedish statistics. The econometric domain absorbed most of my research activity for many years to come. After the dissertation I pursued time series analysis in several articles between 1938 and 1947 and several chapters in my econometric treatise on demand analysis [18], often with emphasis on econometric applications, but without notable innovations. In the introduction to [13], I reported extensive simulation experiments and elaborate graphs on the three main types of stochastic processes; I am told that a pirate edition of the introduction surfaced at Harvard!

2.2. 1938–1952. Econometric Analysis of Consumer Demand

My task in the demand project solicited by the government committee was to measure price and income elasticities of demand, i.e. the sensitivity of consumer demand with respect to changes in commodity prices and consumer income. Rather extensive data were available: time series on demand and on commodity prices, cross-sectional data on family expenditure and family income. The price and income elasticities of food, clothing and other budget items were calculated, using statistical methods to be discussed presently, and the results were published by the government in 1940.

The demand project was initiated when the military build-up before World War II was in full swing, and the war had begun before the results were published. The government had evidently planned the demand project to prepare for a rationing system during the war. Lars Juréen, my assistant in the demand project, took up an appointment in the Food Rationing Agency. In independent studies Juréen showed that our results, with due account for rationing restrictions, were in accordance with consumer behaviour during the war.

As to statistical method, the consumer demand project threw me headlong into a controversial problem area: the choice of regression. In his doctoral thesis of 1906, the Danish economist Mackeprang, posing the problem of how to extract the price elasticity of coffee from time series on coffee price p_t and coffee demand d_t, asked whether one should choose the regression d_t on p_t or the regression of p_t on d_t. The conclusion of his thesis gave the unviable answer: both. Ragnar Frisch in 1934 emphasized that Mackeprang's question had received no answer of general value. Arguing that there is symmetry between the two regressions and between the two variables, Frisch proposed a compromise: diagonal regression. In 1938 Frisch and Haavelmo applied diagonal regression to Norwegian household data to extract the income elasticity of milk demand, and found the elasticity to be 0.80 or more.

From the Swedish household statistics income elasticities could be extracted for food, clothing and other budget items; family income is the sum of the expenditure on the various budget items, including savings. The income elasticity of total expenditure is 1: hence if we consider the income elasticities of the various budget items, the weighted average of the elasticities is also 1. Using this argument in a comparison of the three regressions at issue, I showed that the weighted average of the income elasticities is approximately equal to 1 when the elasticities are estimated by the ordinary least squares (OLS) regression of expenditure on income, whereas the elasticities are grossly overestimated by the diagonal regression, and even more so by the regression of income on expenditure; see [16], p. 218.

The choice of regression was a marshland of diffuse and more or less controversial problems. As to the dangers of too few or too many explanatory variables in a multiple OLS regression, say

$$y = b_0 + b_1 x_1 + b_2 x_2 + \cdots + b_h x_h + e, \tag{4}$$

I illustrated the passage through this Scylla and Charybdis by the following two contrasting theorems ([16], pp. 189 and 194).

(i) Pessimistic theorem. Letting any number be arbitrarily prescribed, say -87, a variable x_{h+1} can be constructed so that when introduced into (4) the coefficient b_1 will take the value -87.

(ii) Optimistic theorem. If the residual e has a small standard deviation, say θ, and its correlations with the x_i-variables are near θ, say $-\theta < r(e, x_i) < \theta$, its biassing effect on the regression coefficients b_i is small of second order, less than $K\theta^2$ where K is constant.

Then there was the question of causal interpretation of OLS regressions. However, Bertrand Russell's denunciation of causality as a scientific notion had had tremendous impact, and throughout the 1940s causal arguments were still practically taboo in the socioeconomic and behavioural sciences. In the early 1950s I began to place emphasis on the operative use of OLS regressions for causal purposes.

In the 1940s the analysis of consumer demand bristled with open problems that could be explored by utility theory combined with calculus and OLS regression. Sweden remained a quiet corner during the war, and in 1942 as Professor of Statistics at Uppsala I could begin to explore demand analysis and related problems both extensively and in depth. Problems in this area were treated by students and associates in my statistical workshop at Uppsala, and after the war in seminar series abroad. The material was brought together in *Demand Analysis, A Study in Econometrics* [16], an exposition of demand analysis that integrates utility theory, probability theory, statistical method, and applications. I was the author of the first four parts, while Part V, written jointly with Lars Juréen, draws applications from my 1940 booklet and from Juréen's papers published during and after the war.

After the war there was a gold rush in macroeconomic model building in econometrics. Several chapters in the [16] deal with demand analysis from the point of view of multirelational models. This theme carries us on to a dramatic aspect of the choice of regression, namely its *raison d'être* relative to ML methods.

2.3. 1945–1970. Econometrics: Path Models with Directly Observed Variables

2.3.1. With Tinbergen's macroeconomic models, published in the late 1930s, a new era begins in econometrics. Taking the variables to be measured as deviations from their mean, let (5a) be the structural form, and (5b) the reduced form of a Tinbergen model.

$$\text{SF}: y_t = By_t + Gz_t + d_t \tag{5a}$$

$$\text{RF}: y_t = [I - B]^{-1} Gz_t + e_t. \tag{5b}$$

The SF contains the behavioural relations and identities that define the model. Key features of Tinbergen's approach are: (i) the parameter estimates B, G which are obtained by OLS regression applied to the SF; (ii) the cyclical properties of the endogenous variables y_t inferred from the RF.

As noted in my thesis [10], Tinbergen's models are *recursive* in the sense that the transformation from SF to RF takes the form of consecutive substitutions of the endogenous variables y_t in terms of the predetermined variables z_t. Thus in a recursive model the relations of SF can be ordered so that matrix B is subdiagonal.

Trygve Haavelmo's seminal works of 1943 and 1949 launched *simultaneous equations* as a general form of multirelational models. Without reference to Tinbergen he dismissed OLS regression as inconsistent when applied to simultaneous equations and instead recommended estimation by ML methods.

2.3.2. Having used OLS regression extensively in earlier work, I felt disturbed by Haavelmo's wholesale dismissal of it. A consumer demand relation is behavioural, and can be conceived of as belonging to the SF of a system of simultaneous equations: if OLS were always inconsistent when applied to simultaneous equations it would be inconsistent when applied to a consumer demand relation.

Spurred on by Haavelmo's rejection of OLS regression, Bentzel and I in 1946 first distinguished whether a system of simultaneous equations is recursive or non-recursive. We then showed that on the classical assumptions of multivariate normal distributions, ML estimation of recursive systems gives numerically the same parameter estimates as OLS regression. So far so good.

Sewall Wright's path models in genetics, as early as 1934, serve purposes of causal analysis. The early 1950s saw a gradual comeback of causal notions. As reviewed in 1981 by Bernert this comeback gained its impetus from several independent but almost simultaneous publications by Lazarsfeld, Simon, and myself [16]; see also Lazarfeld's chapter in [4].

Simultaneous equations were often called path models. New names for recursive and non-recursive systems were causal chain (CCh systems) and interdependent (ID) systems.

The rationale of simultaneous equations/ID systems was established under the "errors in equations" scheme, combined with multivariate distributional assumptions, by Koopmans in 1950. On the applied side the emphasis was on forecasting, and the ensuing multitude of macroeconomic models developed rapidly to meet the ever-increasing demand. At the same time the rationale was under debate; reference is made to the prestigious 1960 symposium "Simultaneous Equations: Any Verdict Yet?" where some serious doubt was voiced (1960). In contrast, CCh systems have a clear-cut rationale, with both SF and RF designed for operative use [11], [18].

In my contributions to the debate I voiced concern about the ID version of "errors in equations," the residual-free relations being assumed to be functionally exact and at the same time reversible. Reversibility in a causal-predictive sense is needed when transforming ID systems from RF to SF. In the real world, however, causal reversibility is at best a rare phenomenon. Controlled experiments, as is well known, are reversible only under ideal limiting conditions.

2.3.3. For a fresh start on relations intended for predictive purposes, I left causality aside and asked myself what could be achieved with a predictive relation like (4); cf. my chapter in the Festschrift for Harald Cramér [12].

The joint multivariate distribution assumptions were also to be left aside, and it is only assumed that the systematic part of the relation is the corresponding conditional expectation. Thus in (4) I assumed (denoting the theoretical parameters by Greek letters):

$$E(y|x_1,\ldots,x_h)=\beta_0+\beta_1x_1+\cdots+\beta_hx_h, \tag{6}$$

and called (6) the *predictor specification* of (4). Immediate implications of the predictor specification (6) are the results

$$E(\varepsilon|x_1,\ldots,x_h)=0 \tag{7a}$$

$$E(\varepsilon x_i)=r(\varepsilon,x_i)=0; \qquad i=1,\ldots,h. \tag{7b}$$

Hence, under mild supplementary assumptions, the OLS regression estimates b_i of the β_i's are consistent in the large-sample sense.

The extension of predictor specification to multirelational models is straightforward, and leads to an array of new insights.

Assuming that the RF of a CCh system (5a, b) allows predictor specification, (8b), then the SF will also allow predictor specification, (8a):

$$E(y|y,z)=\beta y+\Gamma z; \tag{8a}$$

$$E(y|z)=[I-\beta]^{-1}\Gamma z. \tag{8b}$$

Assuming that the RF of an ID system (5a, b) allows predictor specification, (9b), the SF will not allow predictor specification, (8a):

$$E(y|y,z) \neq \beta y + \Gamma z \tag{9a}$$

$$E(y|z) = [I-\beta]^{-1}\Gamma z. \tag{9b}$$

In contrast to functionally exact relations, conditional expectations are not reversible; hence the crucial implication (9a).

As shown by Haavelmo in 1943 by other arguments, (9a) implies that OLS regression is inconsistent when applied to the SF of an ID system.

REID (reformulated ID) systems. Adopting the classical assumption of an ID system (5a, b) that all residuals ε_i of the RF are uncorrelated with all redetermined variables z, we substitute the explanatory (right-hand) endogenous variables y in the SF in accordance with (10a, b):

$$y = \eta^* + \varepsilon \tag{10a}$$

$$\eta^* = E(y|z). \tag{10b}$$

The ensuing REID system has the same parameters β, Γ, and takes the form:

$$y = \beta\eta^* + \Gamma z + \varepsilon.$$

Thus, the explanatory endogenous variables in the SF of the REID system are not the observed variables y but their expected values η^*.

The crucial point is that the SF and RF of REID systems have the same residual ε. Hence both forms allow predictor specification:

$$\text{SF}: E(y|\eta^*, z) = \beta\eta^* + \Gamma z; \tag{11a}$$

$$\text{RF}: E(y|z) = [I-\beta]^{-1}\Gamma z. \tag{11b}$$

The fix-point (FP) method [1], [15], [17]. By (11a) we are invited to estimate the SF by OLS regression, as in CCh systems (5) and (8), but there is the snag that η^* is unknown. The FP method meets this difficulty by an iterative algorithm that substitutes in the sth step η^* in (11a) by its proxy estimate obtained in step $s-1$, and applies OLS regression to obtain proxies for β and Γ; these last proxies then give the proxy for η^* in step s either from (11a) or (11b). As $s \to \infty$ the limiting FP estimates for β, Γ, η^* are denoted B, G, y^*.

In overidentified ID systems the classical assumptions on residual zero correlations are more numerous than the parameters β, Γ to be estimated. The FP algorithm needs only the predictor specification (11a), which implies only as many residual zero correlations as there are parameters β, Γ to estimate. The REID system generalized by (11a) is the corresponding GEID system. As to the RF of GEID systems, the unrestricted predictor specification (11b) of REID systems takes the restricted form:

$$E\left(y|[I-\beta]^{-1}\Gamma z\right) = [I-\beta]^{-1}\Gamma z. \tag{12}$$

The FP family of methods. The key feature of the FP method is that its OLS regressions stay in the SF, and avoid the collinearities that mar the RF. This advantage is the more pronounced the larger the system.

The FP algorithm has been modified in several ways, for example to improve the rate and speed of convergence, without changing the limiting FP estimates, or to extend the scope of the algorithm, or to reduce the standard errors of the FP parameter estimates [17].

2.3.4. *Applications*. As shown in comparisons against other methods for the estimation of ID systems, the performance of the FP family of methods in applications to real-world models and data is distinctly superior as regards prediction accuracy. As to parameter accuracy the comparisons based on real-world models and data are more or less diffuse, inasmuch as there exist no "true parameters" as targets for the comparisons [17].

In applications to simulated data for ID systems generated under the classical distribution assumptions, the superiority of the FP methods in prediction accuracy is maintained, and the lead increases with the size of the system. The simulation experiments further show that as regards parameter accuracy the FP methods are superior for GEID systems; for REID systems they are sometimes superior, sometimes inferior.

2.3.5. *Prediction-oriented versus parameter-oriented estimation*. This dualism shows up clearly in the simulation experiments referred to. The classical distribution assumptions of controlled experiments with nonrandom stimuli is a realm where optimum LS prediction accuracy gives the same parameter estimates as optimum ML parameter accuracy. Outside this narrow realm there is a choice: optimum parameter accuracy does not constitute optimum prediction accuracy, and conversely. To put it in other words there is a trade-off between parameter and prediction accuracy.

The dualism can be seen as a difference in loss functions. In prediction-oriented estimation the loss is assumed to be proportional to the squared prediction error. Thus in the prediction (3) the errors in the parameters b_i are weighted by the variables x_i. This weighting is removed when minimizing the variance of the estimated parameters; hence the difference between prediction-oriented and parameter-oriented estimation methods. For example, Aitken's "generalized least squares method" minimizes the parameter variance, but the variance of the ensuing prediction errors is not minimized.

2.4. 1966–. Systems Analysis: Path Models with Latent Variables

2.4.1. Figure 1 gives a two-dimensional overview of model building and statistical method. Along axis *OE* the models and methods are purely empirical and descriptive, along axis *OT* purely theoretical and explanatory-predictive. The rectangular graph at the upper left illustrates the twofold dichotomy of experimental versus non-experimental data, and descriptive versus explanatory problems, models and methods. With bands for intermediate cases the graph looks like a Scandinavian flag, and is the

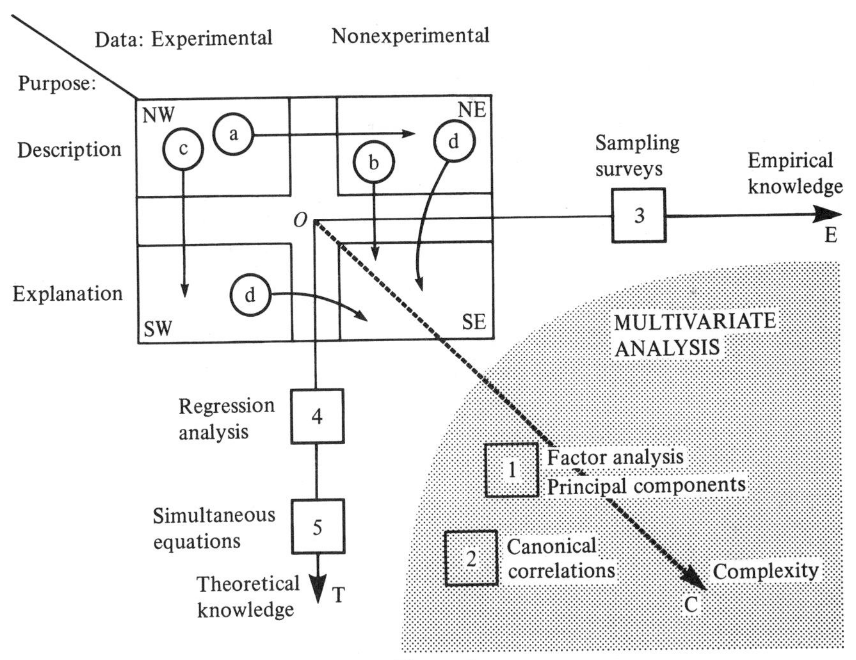

Figure 1

"flag table" which I have used extensively when teaching the historical aspects of statistical method. The north-western quadrant, comprising games of chance, is the area where probability theory was born in the seventeenth century. Arrow (a) marks the energetic evolution of public socio-economic statistics during the "era of enthusiasm 1820–1860." Arrow (b) marks the advent and breakthrough 1880–1915 of the "English school of statistics"; the extension of statistical analysis to data in anthropology and the natural sciences revealed stable empirical regularities that incited novel statistical methods, including correlation and regression analysis. Arrow (c) marks R. A. Fisher's epochmaking work, from 1915 to 1940, on the statistical methods of controlled experiments, accompanied by a bold increase in the level of optimal accuracy possible in statistical inference. The two arrows (d) mark the fact that developments in the explanatory—non-experimental south-eastern quadrant have origins both in the descriptive–non-experimental north-eastern quadrant and the experimental-explanatory south-western quadrant.

In the broad perspective $OE - OT$ the shaded area denotes *multivariate analysis*. The two squares refer to (1) factor analysis and principal components, and (2) canonical correlations. Multivariate analysis has come of age only recently—the 1958 book by T.W. Anderson is the first with this title—and has emerged as an area separate from the developments along axes OE and OT, such as sampling survey methods (3), regression analysis (4) and simultaneous equations (5).

Factor analysis is the principal statistical method of psychometrics, and a large literature reports its applications in psychology and education. Factor analysis works with latent variables indirectly observed by multiple indicators; as illustrated in Figure 1 the factor analysis of psychometrics and the path model with directly observed variables of econometrics are separate avenues of scientific evolution. O. T. Duncan in his seminal 1964 paper initiated a merger of the two avenues. The ensuing model, called *path models with latent variables*, rapidly gained momentum in sociology through the 1960s. This posed new statistical problems, and an array of ad hoc methods were devised for the purpose. Jöreskog in 1967 was the first to implement an ML algorithm for operative estimation of factor models, and very soon followed his ML-LISREL algorithm for general estimation of path models with latent variables indirectly observed by manifest indicators; cf. [7], Part I.

2.4.2. As a by-product of the FP method, I had designed in 1966 an algorithm for assessing principal components by an iterative sequence of simple OLS regressions, and another algorithm for assessing Hotelling's canonical correlations by an iteration of multiple OLS regressions [14]. The conceptual-theoretical design of a path model with latent variables is given in its *arrow scheme*, which shows *outer relations* between the latent variables and their manifest indicators (LVs and MVs), and *inner relations* between the LVs. In 1971, having seen Jöreskog's LISREL approach to path models with latent variables, it struck me that it might be possible to estimate models with the same arrow scheme by an appropriate generalization of my LS algorithms for principal components and canonical correlations. The extension involved two crucial steps, namely from two to three LVs and corresponding blocks of indicators, and from one to two inner relations. Once these steps were taken, the road to an iterative LS algorithm of general scope for estimation of path models with latent variables observed by multiple indicators was straightforward; cf. [7], Part II.

Now some comments on the ensuing estimation method, called the basic design of PLS (partial least squares) soft modelling.

(i) PLS soft modelling takes an intermediate position between data analysis and traditional modelling based on the "hard" assumption that the observables are jointly ruled by a specified probability distribution. The PLS approach is distribution-free, using data observed over time or a cross section, say N cases of observation.

(ii) The formal definition of a soft model can be written down directly from the arrow scheme, and so can the corresponding PLS algorithm. Being a sequence of linear operations and OLS regressions, the PLS algorithm is easy and speedy on the computer: "instant estimation." The first stage of the PLS algorithm is iterative, and provides explicit estimation of each LV as a weighted aggregate of its indicators. The second stage estimates the inner and outer relations by corresponding OLS regressions, and thanks to the explicit estimation of the LVs no identification problems arise. The location parameters of LVs and inner and outer relations are ignored in the two first stages, and are estimated in the third stage.

(iii) As estimation methods of the same type of model, path models with latent variables, I see LISREL and PLS as complementary rather than competitive approaches. LISREL is parameter-oriented, as are ML methods in general, and aims at optimal parameter accuracy. PLS is prediction-oriented, gives predictions that are optimal in the LS sense, and rests content with consistent parameter estimates (consistency in the qualified sense of consistency at large). To repeat from Section 2.3.5, optimal parameter accuracy does not provide optimal prediction accuracy, and conversely.

(iv) Soft modelling allows several modes of predictive inference: (a) prediction of each LV in terms of its indicators; (b) prediction of each endogenous indicator in terms of its LV; (c) each inner relation predicts an endogenous LV in terms of other LVs; (d) by substitution of the LV in (b), using (c), prediction of the indicators of an endogenous LV in terms of other LVs; (e) by further substitution, using (a), prediction in terms of the indicators of other LVs.

(v) For evaluation of the estimated model the LISREL approach uses classical ML criteria, assuming the data to constitute N independent observations of the hypothetical joint probability distribution; typically a likelihood-ratio test on the validity (significance, relevance) of the model, and classical standard errors and confidence intervals for the estimated parameters.

As to model evaluation in the PLS approach, the various modes (iv), (a)–(e) of predictive inference are tested for predictive relevance by straightforward adaptation of the test proposed in 1974 by Stone and Geisser; that is, for each prediction R^2 is evaluated without loss of degrees of freedom. As to the accuracy of PLS parameter estimates, standard errors are obtained by Tukey's jackknife method.

(vi) LS prediction is an optimization principle with a standing of its own relative to ML optimization of parameter accuracy. ML methodology has a well-established framework. An LS framework is beginning to emerge on the basis of predictor specification and LS estimation; Stone–Geisser testing, and jackknife standard errors.

2.4.3. *Applications.* The reported applications of PLS soft modelling range widely from natural science and medicine, often with reproducible data, across to the non-reproducible data of socioeconomic, behavioural and political sciences. Typically, the applications belong under systems analysis in terms of structure vs. performance, with structure modelled by one or more LVs observed by multiple indicators, and similarly for performance. Note that it does not matter whether the observables are scalar or ordinal (or categorical, as is the case in multidimensional contingency tables); since PLS is distribution-free the PLS algorithm remains the same. The applications give evidence of the flexibility of PLS soft modelling, the investigator being free to design the arrow scheme in line with previous knowledge and/or new ideas, and in particular the design of the path of inner relations between the LVs, and the choice of indicators for the various LVs.

The flexibility of the basic design of PLS soft modelling is extended by several generalizations. Among these hierarchic structure of the LVs; are LVs in two or more dimensions (special cases: principal components and canonical correlations of second or higher order); hierarchic structure of the LVs; both latent and manifest variables in the inner relations; feedbacks (interdependencies) in the inner relations; indicators observed both over time and space.

PLS soft modelling is at an early stage of evolution in both theory and practice. From the reported applications it emerges that PLS soft models are being used for the first time in quantitative analysis in several fields of systems analysis where traditional statistical methods have not been applied because of the complexity of the system. In such virgin areas, fresh quantitative knowledge is provided already by the rough order of magnitude of the correlation between the LVs of structure and performance, or of a beta coefficient of an inner relation to the first decimal place.

2.4.4. *Aspiration levels of accuracy*. The trade-off between parameter and prediction accuracy is at the heart of the matter; cf. Section 2.3.5. LISREL aims at optimal accuracy of the parameters, including the distributional parameters of each LV, but does not estimate (predict) the N case values of each LV. PLS provides explicit, albeit deliberately approximate, prediction of the N case values of each LV; to repeat, the ensuing parameter estimates are consistent at large, but do not aim at optimal accuracy. Accordingly, the various PLS modes of predictive inference outlined in Section 2.4.2 (iv), (a)–(e) work at different levels of accuracy. When an LV has only a few indicators the explicit PLS estimation of its case values is very crude; hence the ensuing correlation between the LVs often is grossly underestimated. In contrast, the parameters of the outer relations in general are overestimated by PLS, whereas the ensuing coefficients of substitutive prediction of Section 2.4.2 (iv) (d)–(e) are nearly consistent.

The more numerous the indicators of an LV (i.e. the larger its block of indicators), the more accurate are the PLS estimates of its N case values. More precisely, the PLS estimates of LVs and of inner and outer relations are *consistent at large* in the sense that they become consistent in the limit as the block sizes increase indefinitely.

(i) The forte of LISREL lies in accurate parameter estimation in simple or fairly simple models. In models with some few LVs and inner relations the interpretation and operative use of the model and its parameters is simple and clearcut. In the passage to complex models with many inner relations and parameters, PLS soft modelling comes to the fore. For one thing, in the operative use of the model the emphasis shifts from prediction in terms of individual parameters to the integrated use of the parameters in the various modes of prediction from the model (cf. Section 2.4.2 (iv), (a)–(e)). Second, the LISREL computer work increases rapidly with the complexity of the model, and becomes intractable when the model has more than, say, 50 or 60 parameters, whereas PLS in recent reports is successfully applied to models with more than 100 parameters, using less than one minute of computer time.

3. Reflections on Research as a Creative Process

3.1. Early Opportunities in Research

When I first came from a small town to the University of Stockholm in 1927 for my undergraduate studies, I had little or no knowledge of the academic milieu, let alone of research. Having obtained good school marks in mathematics and natural science, I enrolled for mathematics in my first year, and then joined the audience of Harald Cramér, the young holder of the new chair in insurance mathematics and mathematical statistics. To belong to Cramér's first group of students was good luck, an advantage that simply cannot be exaggerated.

In the early 1930s Sweden was suffering from the great depression. After receiving my B.Sc. degree in 1930 I was interested in earning some money, and on Cramér's recommendation I was appointed by the Fylgia Insurance Co. to compile Swedish rainfall statistics and construct a tariff for rain insurance. The ensuing report was accepted by Cramér as my licentiate thesis in 1933, and for several summers I was in charge of Fylgia's rain insurance, one of the branches under the charge of Mac Hall, its managing director. I can never sufficiently express my admiration for him and my gratitude for his guidance in practical affairs.

The organization of the Swedish university system follows the German pattern. J. D. Bernal in his 1934 monograph *The Social Function of Science* described the advent of the modern university in Germany after the Napoleonic wars, with its systematic coordination of teaching and research, and its emphasis on both fundamental research and research directed towards social progress. Bernal describes how the body of human knowledge expands through research. Advances at the frontier of research have different degrees of relevance; important advances open up new vistas and attract followers who develop the virgin fields. T. S. Kuhn in 1962 describes research as a dichotomy between normal science and scientific revolutions; but this is too crude. At the research frontier it is a scientist's normal ambition to make an important advance, thereby provoking a "gold rush" of followers to explore, develop and consolidate the new territory.

The definition of probability had been fruitlessly debated since the birth of probability theory in the seventeenth century. In establishing probability theory as a branch of rigorous mathematics in 1933, André Kolmogorov achieved a genuine scientific revolution. Harald Cramér, with a background in number theory and familiar with Kolmogorov's earlier work, had already presented Kolmogorov's theory in a graduate course on stochastic processes in 1933–34. Cramér posed an array of problems on the main types of stochastic processes. Attracted by the theory and applications of stationary time series, I got a flying start in this field, and in due course was able to take up problems of my own for my doctoral thesis in 1938 [10].

Opportunities for research are largely determined by a scholar's academic environment and background; these opportunities may arise at a distance from the current research frontier, or in an active sector of it. The ensuing research may be called *opportunity research*, an apt term for which I am indebted to Irma Adelman. Although my mathematical background was modest, I was extremely fortunate in my opportunity research during my graduate studies in being an early pupil of Harald Cramér, and I wish to pay tribute to the stimulating atmosphere at his institute, which in the 1930s rapidly developed into a lively research centre with many visitors from abroad. Among them was Willy Feller, who spent two or three years there on his way from Göttingen to Princeton via Copenhagen and Stockholm.

As I turned to econometric analysis of consumer demand in 1938, the Goddess of Good Opportunity again smiled upon me. First, the problem area was of great importance in theory and practice. Second, I entered directly into a very active sector of the research frontier: the choice of regression (see Section 2.2).

When I came to Uppsala as Professor of Statistics in 1942, I decided to concentrate my research on econometrics. In hindsight, three aspects of econometrics combined to attract my interest; the subject was a melting pot of new developments in need of (1) applied statistics, with emphasis on substantive analysis; (2) a synthesis of economic theory, probability theory, statistical method, and applied statistics; (3) research on open and/or controversial problems. A beginning under (1) had been made with my 1940 government booklet. With a tenured position at Uppsala I could take up research on (2) and (3). The Statistics Institute had a limited staff: one professor, one half-time assistant, one half-time secretary. Then in 1945 the Hiroshima bomb awakened the Swedish government to a forceful expansion of the universities. Thus in 1946 the staff of the Uppsala Statistics Institute suddenly increased from two to twelve. There was a further increase under the government's aid program for Baltic refugees; at one time the staff included ten Esthonian refugees.

3.2. Self-initiated Research

With these added resources, my research under (1)–(3) was summed up in [18]. If we distinguish between opportunity research, gold-rush participation and self-initiated research, I see my first ten Uppsala years as a transition from opportunity research to self-initiated research (without participation in the ML gold rush). Since then, most of my research has been self-initiated. Let me give some personal reflections on my attitude to such research, and then add some more general comments.

(i) I wish to pay tribute to two main incentives and sources of inspiration. First, applied statistics: what has usually given me the decisive hint when dealing with controversial problems is the study of substantive aspects of earlier applied work, and less often the study of earlier theoretical work. A typical case in point is the third axiom in my axiomatic exposition of utility theory; another is the ensuing integrability condition of preference fields ([18], pp. 82 and 90 ff.).

(ii) Secondly, and more precisely, my argument on controversial problems has often taken a dialectical form: the obscure issue is explored in some depth, first to determine the parting of the ways, next to form thesis and antithesis, and then to set up a synthesis for resolution of the problem. Cases in point are the choice of regression (cf. Section 2.2), and REID systems versus classical ID systems (cf. Section 2.3.3).

(iii) Some of my self-initiated research breaks away from the mainstream or from current views in the literature. Cases in point are my FP and PLS algorithms, which are LS and therefore not in the ML mainstream (cf. Sections 2.3.3 and 2.4.2); and also my definitions of the notions of model and cause-effect relationships (cf. Section 3.3).

At a distance from well-trodden paths, I have the feeling of groping in the dark; a false step may be dangerous. I proceed slowly, and report preliminary results in an array of progress papers; I reap the fruits of collaboration not so much from discussion while writing a preliminary paper, but far more from discussion when it has been released.

(iv) I should like to state, as a general conjecture, my experience of correct and false steps. Having taken a correct step, it is easier to make the next correct step forward; but a false step leads nowhere, and does not give a lead for the subsequent step.

(v) As I see it, point (i) above is also of general scope. I have often been amazed to find that when focussing on the substantive aspects of a real-world problem you need only scratch the surface to get important clues for theoretical and applied research. To put it otherwise, owing to its overexpansion the marginal productivity of purely theoretical research is low relative to that of substantive empirical research. To put it in terms of research evaluations: in current practice marks are too high for purely theoretical research, and far too low for joint theoretical–empirical research.

(vi) Among experienced researchers working at the periphery or outside the mainstream, there is some consensus that established learned journals adhere to the mainstream; their referees do not have the function of evaluating contributions that break away from this mainstream. Time-consuming and futile debates with referees can be avoided by going elsewhere to publish dissident research, say in Festschrifts or other occasional volumes.

This pragmatic game I have played with my papers on the philosophy of science, and my early papers on PLS; cf. [12], [14].

3.2.1. In the 1950s and 1960s my research interest broadened from econometrics to non-experimental analysis in general, and to the philosophy of science. Again my interest was attracted by notions that were fundamental and at the same time obscure or somewhat controversial. Cases in point are the notions of *model* and *causality*. Authors concerned with these much-debated issues often take a premature position on a controversial problem; for me only a gradual approach was possible.

Suppes in 1960 reviewed 16 different definitions of the scientific notion of model. My definition [16] is composed of four elements, and illustrated as follows:

$$\boxed{T \Leftrightarrow E}$$

The rectangle is the *frame of reference*, which describes the problem under investigation in ordinary language, and links up with earlier related work; T is the *theoretical content* of the model; E is its *empirical content*; the double arrow indicates that when estimating and testing the model T and E are subject to a *matching process*. Here are some comments for later reference.

(i) In the philosophy of science it is often argued that theoretical and empirical knowledge cannot be kept strictly apart. The frame of reference is an inseparable composite of theoretical and empirical knowledge, but thanks to it T and E are kept strictly apart *within the model*.

(ii) The frame of reference is the distinctive feature of my definition of model. *Model for knowledge* (French: modèle de connaissance; German: Erkenntnismodell; Swedish: kunskapsmodell) is my name for such models.

(iii) Controlled experiments are special cases of models for knowledge. T includes the null hypothesis of the model, and E includes the recorded outcome of the experiment.

(iv) If a model for knowledge, say M, is part of a research program (as defined by Imre Lakatos in 1970), this affiliation should be spelled out in the frame of reference of M.

(v) "Model for knowledge" is a notion of general scope: it is my contention that any human knowledge can be represented by such a model.

(vi) For example, the relationship of cause and effect can be defined by a model for knowledge [16]. The definition is in two steps: (1) the stimulus-response relation in a controlled experiment is a special case of cause-effect relations; (2) in non-experimental analysis, a cause-effect relation is defined by assuming the observed relation at issue to be the outcome of a fictitious controlled experiment.

3.2.2. The Danish psychologist Hoeffding [6] described and gave examples of what may be called the *Hoeffding effect*: a scientist runs into emotional turmoil and gives up research work for a long time; when the turmoil is over he returns to the scientific workshop and the interrupted research project; then he finds that the relief from the turmoil brings an appreciable increase in the efficiency of his research. For a case in point, see Peter Whittle's gem of an article in [4].

In the 1960s I experienced what I interpret as a Hoeffding effect, although the turmoil had been intellectual rather than emotional. Through the 1950s I had discussed causal chain (CCh) system versus interdependent (ID) systems in several papers, emphasizing the clear-cut causal interpretation and operative use of the structural form (SF) of CCh systems on the one hand, and the obscurities around the SF of ID systems on the other (cf. Section 2.3.2). Observing over the years that the causal arguments had little or no impact, an intellectual turmoil built up within me in the late 1950s. Then when serving as Visiting Professor at Columbia University in 1958–59, I gave up. I decided to push the causal arguments on CCh and ID systems aside, and make a new start on the basis of predictor specification (cf. Section 2.3.3). The new approach led very soon to the FP method (cf. Section 2.4) and then to PLS soft modelling (cf. Section 2.5).

3.3. The Experience of Research

On reading Sir Peter Medawar's *Advice to a Young Scientist* [8] I was struck by the fresh and concise exposition of important features in his broad and rich experience. With admiration I quote some of his points:

(i) Science is the *art of the soluble*.

(ii) The research process is *imaginative guesswork*, usually not a logical deduction from earlier knowledge.

(iii) The criterion for relevance and validity of a new insight is that it leads to further insights.

(iv) The characteristic feature of most successful research workshops is *synergy* in individual and team work—open discussion and exchanges of thought on current projects.

(v) Devoted researchers have all sorts of temperaments—the sole feature in common is that for them research is the only kind of life that will do.

In elaborating this last point, Medawar pays tribute to the scientific life—an exciting, rather passionate, very demanding and sometimes exhausting occupation. Medawar gives advice on what lifestyle is helpful for research, and I can only agree, without quoting in detail: a modest, quiet and untroubled life; long working hours; exercise in the open air to promote stamina and health. The working scientist never thinks of himself as old, and as long as circumstances allow him to continue with research he enjoys the young scientist's privilege of feeling himself born anew every morning.

This obsession is tough on wife and children, and calls for their generous tolerance.

My excitement at Medawar's essay is combined with delight since the above and many other points are in line with my own experience. Let me adduce some examples of his points (ii)–(iv).

In applied statistics the problem under analysis is partly theoretical, partly empirical, and the imaginative guesswork is a process of matching theoretical and empirical knowledge. As to the choice of regression (cf. Section 2.2), my first argument in favour of OLS regression of expenditure on income was the matching of a theoretical weighted average with the corresponding empirical average.

Note the affinity between Medawar's point (iii) and mine on groping in the dark (cf. Section 3.2 (iv)). Scores of joint papers of mine reflect synergy in Medawar's sense. Specific reference is made to my seminar workshops on FP (cf. Section 2.4), and PLS (cf. Section 2.5 and [15], [17]). An early instance is my 1940 Swedish research report on demand analysis, with Lars Juréen as my assistant. To give an example, I was impressed by the regularities of the statistical data, in particular the quarterly data that Juréen had compiled on food quantities sold at the cooperative stores in Stockholm. On examining the data, Juréen observed that one of the quarters was not quite in line with the regular yearly cycle. Juréen said, "they must have forgotten to include one of the cooperative stores," and on checking at the cooperative data centre, he retrieved the missing item, bringing the data back to smooth regularity.

Medawar gives fine examples of *serendipity*, sheer good luck, in the research process. At one of my first seminars on the FP method, at the University of North Carolina in 1964, a participant asked whether my iterative algorithm was related to an iterative procedure he had devised to evaluate the skill of farmers from yearly data on their farm profits. This gave me the clue for computing principal components by an iterative procedure, cf. [14]. This case of serendipity in turn led me to an iterative procedure for the computation of Hotelling's canonical correlations [14].

I see the birth of PLS soft modelling as another case of serendipity. In 1971 when I saw Jöreskog's LISREL algorithm for ML estimation of path models with latent variables, it struck me that the same models could be estimated by suitable elaboration of my two algorithms for principal components and canonical correlations (cf. Section 2.5).

My decomposition theorem on stationary processes was a vehicle for serendipity elsewhere, inasmuch as it was carried over to spectral and operator theory by Kolmogorov and Wiener (cf. Section 2.1). It so happened that the decomposition theorem opened new vistas in the abstract spaces of spectral and operator theory, but in the statistical setting the theorem was near at hand as a synthesis of well-known special cases.

3.4. Further Thoughts on the Research Process

In Medawar's advice to scientists young or old, the quotations of Section 3.2 (i)–(v) apply generally from natural science and medicine across to socio-economic, behavioural and other non-experimental sciences. More specifically, his masterly discussion of psychological aspects of the research process is of general scope in experimental and non-experimental sciences. When it comes to technical matters Medawar's discourse is primarily concerned with natural science, and here his exposition is supported by excellent arguments and illustrations. Some of these arguments do not extend beyond natural science, and in the passages where he refers directly to non-experimental sciences his argument is not always to the point.

Let me begin with a minor point. As to verbal expositions, Medawar warns strongly against reading from a manuscript; I agree, for I tend to fall asleep if a verbal exposition does not appear spontaneous. Medawar recommends the use of a blackboard to illustrate the verbal exposition, and warns against slides.

A blackboard may suffice to illustrate a controlled experiment, with stimulus, response, null hypothesis and counterhypotheses. But in non-experimental sciences the model is often much more elaborate, and both speaker and audience gain when the exposition is illustrated by overhead projection of predesigned transparencies, as is now the widespread practice.

In his chapter "Experiment and Discovery" Medawar distinguishes between four types of experiment: Baconian, Aristotelian, Galilean and Kantian. The first two belong to the historical background of science. In current statistical usage, Galilean experiments are synonymous with controlled experiments, and what Medawar by a far-fetched term calls "Kantian experiment" is synonymous with research on non-reproducible data, also known as non-experimental research. Controlled experiments are Medawar's territory, or in his own words the "hypothetic-deductive" approach in "the Galilean spirit." On p. 85: "Again, a good hypothesis must also have the character of *logical immediacy*, by which I mean that it must be rather specially an explanation of whatever it is that needs to be explained and not an explanation of a great many other phenomena besides." The logical immediacy of a controlled experiment is that of a model for knowledge with a frame of reference that explains what needs to be explained as a prerequisite.

Medawar's "Kantian experiments" belong to the non-experimental sciences. This is an enormous domain, but Medawar is here outside his own field, and his brief discourse gives the reader little or no orientation. There is no mention of the essential distinctions between reproducible and non-reproducible data, or between controlled experiments and non-experimental research. The word "model" does not appear in Medawar's book, and there is no mention of important contemporary model-building developments in socioeconomic, behavioural, political and other non-experimental sciences.

To conclude with a minor point: in his chapter on "Presentation" Medawar recommends that a scientific report should have an introduction, an exposition, and a summary, and warns strongly against a concluding "discussion." This may be fine in reporting controlled experiments. In contrast, non-experimental models often have an array of potential alternatives that would burden the initial frame of reference, and are therefore better relegated to a concluding discussion of what the results suggest about directions for further research.

4. Discussion

The matching of theoretical and empirical knowledge is the main thread of my contributions. Having taken stock of my work, let me turn from the past and comment on the outlook for the future.

PLS soft modelling is designed for the analysis and evaluation of systems indirectly observed by multiple indicators. The recent applications of soft modelling show that PLS has pushed the research frontier forward a little in coping with large and complex systems, their many variables and huge data masses. But only a little. PLS is at an early stage of evolution in theory and practice; the realm of systems to explore is enormous; only a small beginning has been made, and further progress is by no means a matter of routine.

PLS models the systems in terms of structure versus performance, with directly observed indicators for the latent variables of structure and performance. An array of adequate indicators is needed both for structure and performance. When it comes to systems evaluation, the performance indicators are of crucial importance, and this is a territory where it may be difficult to obtain adequate data, and where the investigator must watch out for biasing influences, pitfalls, or worse. School systems are a case in point; the analysis of school systems is an established area of substantive research, whereas the evaluation of school systems lags far behind; there is no consensus in the selection of adequate performance indicators, and evaluation is in rather a mess. Difficulties are increased by the fact that the evaluation of schools and other large systems is partly a political matter, and is on the agenda of public administrators and of political bodies. The larger the system, the more room for political and administrative considerations in the evaluation. And the larger the system, the more room within the system for "enormities," in Arnold Toynbee's sense.

Thus if we turn from schools to universities the problems of system evaluation become even more difficult. "Number of published papers" is the criterion in general use as the sole or dominant indicator of research performance. "Publish or perish" is the researchers' response. Everybody knows that the quality of a research paper is an essential feature of its evaluation, but how to measure quality is a question that is easier to ask than to answer. The precept "publish or perish" thus goes a long way to explain the enormous amount of purely theoretical research in contemporary statistics. Statistical theory has its *raison d'être* in the ensuing substantive applications, but of course it is much less time-consuming to produce a purely theoretical paper than to combine the theory with a substantive application. In consequence, we have an "enormity" in Toynbee's sense. The enormity of pure theoretical research is apparent also in econometrics, psychometrics, and other sciences where theory has branched off from the initial joint body of theory and applications. To counteract such institutional enormities is no easy task. Substantial weight should be given to joint theoretical and applied work. Another weakness of the present evaluation system is that, although follow-up studies of applied research are often very important, they are usually given little or no weight.

Broadening the view from school and university to society as a whole, one may ask whether systems analysis and evaluation can make any relevant contribution in counteracting the many severe problems of our troubled world. Modern society bristles with enormities in Toynbee's sense. Douglass North [9] dismisses much of contemporary economics as a failure since its analysis of inflation, unemployment, and other problems regards the institutions of society as fixed, rather than subject to long-range economic-historical changes. While Toynbee and North give no solutions to the problems of modern society, they do give clues to latent variables essential for the analysis and evaluation of modern society in terms of structure and performance.

Acknowledgement

The present paper was written under a grant from the Stiftung Volkswagenwerk in support of research on PLS soft modelling.

My thanks are due to Professor Irma Adelman, Berkeley; Professor P. Balestra, Geneva; Professor K. G. Jöreskog, Uppsala; J.-B. Lohmöller MA, Munich; Professor W. Meissner, Frankfurt am Main; and Dr. Svante Wold, Umeå, for reading and commenting on the first draft of the paper. As is customary, I wish to emphasize that the responsibility for any errors of fact or interpretation rests with me alone.

Publications and References

[1] Bergström, R. and Wold, H. (1982) *Fix-Point Estimation*. In Applied Statistics and Econometrics, ed. G. Tintner, H. Strecker and F. Féron. Vandenhoeck and Ruprecht, Göttingen.

[2] Bernert, C. (1982) The entry of causal analysis to American sociology. *British J. Sociol.* To appear.

[3] BORCH, K. (1981) Additivity and Swiss premiums. *Mitt. Verein. Schweiz. Versicherungssmath.* 19–25.

[4] DALENIUS, T., KARLSSON, G. and MALMQUIST, S. (Eds) (1969) *Scientists at Work; in Honor of Herman Wold.* Almqvist and Wiksell, Stockholm.

[5] HELSON, H. (1964) *Lectures on Invariant Subspaces.* Academic Press, New York.

[6] HOEFFDING, H. (1982) *Psychology Based on Experience* (in Danish).

[7] JÖRESKOG, K. G. and WOLD, H. (Eds) *Systems under Indirect Observation : Causality* Structure* Prediction* I, II. North-Holland, Amsterdam.

[8] MEDAWAR, P. B. (1979) *Advice to a Young Scientist.* Harper and Row, New York.

[9] NORTH, D. (1981) *Structure and Change in Economic History.* Norton, New York.

[10] WOLD, H. (1938) *A Study in the Analysis of Stationary Time Series.* (2nd edn 1954, with appendix by P. Whittle.) Almqvist and Wiksell, Stockholm.

[11] WOLD, H. (1955) Possibilités et limitations des systèmes à chaîne causale. *Cahiers Sém. Econom. R. Roy* **3**, 81–101. CNRS, Paris.

[12] WOLD, H. (1959) Ends and means in econometric model building: basic considerations reviewed. In *Probability and Statistics, The Harald Cramér Volume*, ed. U. Grenander, Almqvist and Wiksell, Stockholm: Wiley, New York, 355–434.

[13] WOLD, H. (1965) *Bibliography on Time Series and Stochastic Processes.* Oliver and Boyd, Edinburgh.

[14] WOLD, H. (1966) Nonlinear estimation by iterative least squares procedures. In *Research Papers in Statistics: Festschrift for J. Neyman*, ed. F. N. David, Wiley, New York, 411–444.

[15] WOLD, H. (1967) *Forecasting on a Scientific Basis.* Gulbenkian Foundation, Lisbon.

[16] WOLD, H. (1969) Mergers of economics and philosophy of science, A cruise in shallow waters and deep seas. *Synthèse* **20**, 427–482. (Spanish translation: *Metodologia y Critica Economica*, ed. C. Dagum. Fondo de Cultura Economica, Mexico.)

[17] WOLD, H. (Ed) (1980) *The Fix-Point Approach to Interdependent Systems.* North-Holland, Amsterdam.

[18] WOLD, H. and JURÉEN, L. (1952) *Demand Analysis: A Study in Econometrics.* Almqvist and Wiksell, Stockholm: Wiley, New York.

6
STATISTICIANS IN DEMOGRAPHY AND MEDICINE

B. Benjamin

Bernard Benjamin was born in London on 8 March 1910, the youngest of eight children. He was educated at Colfe Grammar School, Lewisham and later read physics at Sir John Cass College, London.

He has had a varied career, and has held the positions of Director of Statistics in the United Kingdom Department of Health up to 1965, Director of Research and Intelligence for the Greater London Council 1965–70, and Director of Statistical Studies at the Civil Service College 1970–73. Having retired from this last post, he was promptly appointed Foundation Professor of Actuarial Science at the City University, London in 1973. He is now Visiting Professor, but acts as Director of the City University Centre for Research in Insurance and Investment, an interdisciplinary unit which he has formed.

Dr. Benjamin has been closely involved in the development of actuarial, demographic and health-related statistics both in the United Kingdom and internationally. He was elected President of the Institute of Actuaries for 1966–68, and President of the Royal Statistical Society for 1970–71, the only person so far to have held both appointments. He has also been Vice-President of the International Statistical Institute for 1973–77. Among his many honours are the Institute of Actuaries' Gold Medal 1975.

Dr. Benjamin has written several books and papers on actuarial and demographic topics. He was married in 1937 and has two daughters. His hobbies are music and painting.

A Statistician in the Public Health Service

B. Benjamin

1. School and Early Professional Training

I left Colfe Grammar School in London in 1928. I had not obtained university scholarships that would sufficiently relieve my parents of the expense of supporting me for three years and so I had to enter the labour market. The world recession of 1930 was already looming ahead and my father's advice was to take any secure job that I could get. I entered for, and passed, the examination for entry as a general grade clerk in the London County Council (LCC) which then governed what is now the inner area of Greater London.

Because I had passed especially well in mathematics, I was allocated to the actuarial office concerned with administering the LCC's own pension fund and I was immediately introduced to the routine computations for the 1928-based valuation of the assets and liabilities of that fund under the direction of two resident actuaries, Thomas Tinner and Sidney Alison. Tinner was the senior and spent much of his spare time (which in those days was significant in quantity) rewriting a well-known textbook of algebra; he would try out a new page on me every morning. Alison was the driving force; a brilliant actuary with an absorbing interest in, and a mastery of, all the natural sciences. He found it difficult to tolerate less agile brains, but he did so, for he was a very kind man. He was to be a constant source of encouragement and inspiration to me.

I had left school with a playful taunt in my ears which I had taken seriously. The physics master had said in my last day, "Get on with the experiment—this is the last thing you will ever know about physics." I determined to prove him wrong, and I promptly registered as an evening student at the Sir John Cass Technical Institute to prepare for an Honours B.Sc. in Physics; the Institute was an internal school of the University of London. Five evenings a week I walked from Westminster Bridge to Aldgate and often stood for four hours in the laboratory doing practical work. I had to interrupt my studies to prepare for an examination to move up into the major establishment of the LCC (the equivalent of the administrative class of the Civil Service) and so I did not obtain my degree until 1933. I then found that there was a surplus supply of physicists. "Well," said Alison "why not become an actuary?" And so I began a long period of spare-time study to qualify. In those days the norm was about seven years; full-time study and more rapid progress was not to come until forty years later.

During my physics studies I had met something called "probable error" which was then being attached to many physical laboratory measurements and Alison had introduced me to David Brunt's "combination of observations," which whetted my appetite to learn more of statistical theory. The actuarial syllabus of the time included only elementary statistical theory. The Institute of Actuaries did not then have its own textbook but used that of Caradog Jones.

2. The Public Health Department, LCC

At about the time when I took this particular part of the examinations, a statistical post in the public health department of the LCC was advertised and as it involved a promotion I applied for it. I was successful, and joined a small section principally engaged in collecting and analysing discharge summaries for all patients passing through the LCC general hospitals [20]. This was in 1936, long before the advent of the National Health Service. The LCC had acquired the former Poor Law hospitals in 1930, and had rehabilitated and developed them into very good general hospitals. Prior to 1948 the LCC was the largest hospital authority in the world, with 93 hospitals of various types [19].

In fact, the discharge summaries were entered onto dual purpose cards suitable for punching and processing on the then conventional Powers–Samas machines; these could sort cards into the categories of each subject column and count them. The procedure was sometimes laborious. For example, measuring average duration of stay meant counting 100's, 10's and units of days, multiplying up by 100, 10 and 1, adding the sums together and dividing by the total number of patients. We were able to give a service to interested clinicians very similar to that of the present Hospital Activity Analysis of the National Health Service [11].

But I was not satisfied. The head of the statistical section was not a professional and had no concept of using statistics for decision making (in those days there were few professionals, even, who had such a concept). He did not offer anything beyond purely descriptive statistics. As a consequence the medical staff did not associate us with the task of statistical analysis; they went elsewhere. I remember being horrified to find that the analyses of the biennial measles epidemic (measles was then a serious disease) was handed over to the London School of Hygiene. When I complained of the absence of any analytical work to relieve the drudgery of descriptive tables, I was rewarded with a heavier load of drudgery.

In 1937 I married and, with the support of my wife, not only continued my actuarial studies but began to interest myself seriously in statistics generally and especially in the statistics of medicine. I read medical textbooks and journals. We had in the office a beautifully bound complete series of *Annual Reports of the Registrar-General* with the graphic letters to the Registrar-General from William Farr. I read them all avidly and I still quote from them. (Farr has remained my inspiration to this day.) I questioned my medical colleagues. I found that in 1937, and for at least two decades after, most medical practitioners took the stern view that no lay statistician could be trusted to analyse the statistics of disease, and should not even be allowed to try. Despite this prevalent attitude, many of them were nevertheless prepared to explain medicine to an earnest young man who wanted to know what it was he was counting, and who was prepared to listen patiently and modestly.

It was then that I became convinced that there existed no such subject as *medical statistics*, any more than there existed chemical statistics, or town-planning statistics, or engineering statistics. Statistical theory did not change with the environment in which it was practised, but I realised that anyone hoping to practise statistics in medicine would have to make the effort of learning the language of medicine. For two more years the statistical section continued to be underused.

3. Wartime Service

Then in 1939 we found ourselves at war. Immediately, those of us in the public health department who were not called directly to the armed forces were decentralised to hospitals near our homes to assist the hospital administrators. It was expected that there would be very large numbers of air-raid casualties. I went on night duty at my local hospital with the nominal task of identifying casualties as they arrived and making a daily listing of these for the central government. But there was a quiet period before the raids began and to relieve the boredom I completed my actuarial studies and also wrote my first paper for a medical journal (eventually published in the *Journal of Hygiene* [1] in 1943). This was concerned with the seasonal variation in the growth of schoolchildren, based on a survey of heights and weights conducted in 1939.

The lull did not last long. When the heavy bombing raids on London began, my nights were occupied in trying to identify casualties as they were unloaded onto trolleys and while the medical staff decided whether the trolleys should be wheeled to the surgical theatre or to the mortuary. It was a deeply shocking experience. I had never seen a dead body before. The first casualty I saw (actually handed into my arms by a police patrol officer) was a small boy with his brains obtruding from a smashed skull. He turned out to be the son of a consultant at another hospital (the family air-raid shelter had received a direct hit), so I had the further traumatic experience of informing the father. In a few minutes I had discovered the full horror of war.

After a few months I was recalled to the LCC's County Hall. The government had decided that as part of its surveillance of wartime morale it needed to have health measurements of the population of London. I was required to set up a rudimentary statistical section to provide an intelligence service to Sir Allen Daley, the Medical Officer of Health for the County of London. I had one statistical clerk and three punched-card operators. The punched-card machines had, foolishly, been left on the top floor of County Hall and had been hit by a small bomb. Somehow they were repaired. Important punched-card records had been bombed too, and we spent days retrieving cards from the surrounding streets—some were found 600 yards away. But, remarkably enough, all were retrieved and repunched onto new cards.

This was the beginning of a very rewarding period of collaboration with Sir Allen who had the essential quality of a first-class administrator–the capacity to assimilate information and to use it. Before I was released to join the RAF (as a radar mechanic) we published a joint paper in the *British Medical Journal* on the effect of war conditions on the prevalence of tuberculosis [18]. Sir Allen could easily have put the paper forward on his own as an official report from the Medical Officer of Health, but he deliberately added my name, partly in appreciation of my work and partly to set the seal of medical approval on my intervention as a lay statistician in medical work.

I must mention an incident which also improved my status at the time. There was a problem of increasing mortality from tuberculosis among mental hospital patients. A special committee was set up to investigate this and the late Professor Major Greenwood of the London School of Hygiene, a famous statistician in medicine, was called in as consultant. Then still a junior figure, I was in awe of the great man. But at the first meeting he deliberately prefaced his remarks by deferring to my authority as the resident statistician. He need not have done so and it was a very courteous gesture. The medical superintendents of the hospitals, who were slightly suspicious of me, were clearly impressed.

4. Growth of the Public Health Statistical Section, LCC

After demobilisation in 1946 I was put in charge of a rehabilitated statistical section. I was still the only professionally qualified statistician, but I now had a mathematical graduate and two senior administrators who were statistically minded and were good organisers of work. I was assisted by several statistical clerks and a much strengthened punched-card unit. I adopted a new approach of going out to seek work rather than waiting for it to come to me. Working on the basis that nothing succeeds like success, I set myself the task of completing one or two quick jobs that were really useful in terms of assisting policy making. In a short while we were exhaustingly, but excitingly, overworked. One of the difficulties of the public service is the lag in the provision of resources. You have to do the job *without* the resources in order to prove that you *need* the resources. Much of my spare time was taken up with work brought home from the office.

I continued to receive great help from Sir Allen Daley. He encouraged me to publish in the medical press either under my own name, or in collaboration with individual members of the medical staff. It was Sir Allen's habit to annotate his agenda for the fortnightly meeting of the Health Committee with requests for information and to send it to me. Against an item "proposal for the opening of a new tuberculosis clinic" I would read "Brief note on present trend of prevalence and mortality from TB in London." Every piece of information was carefully read and absorbed and, when Sir Allen was called upon to speak, it was repeated flawlessly and with great authority. Nothing pleases a professional statistician more than to have his work put to good use. The good statistician does not want to pre-empt the decision making of his manager, but he certainly wishes to be listened to. Sir Allen was a very good listener who acknowledged those who kept him informed. Often he would refer to me publicly by name as the source of the information.

In 1948 the National Health Service was introduced and the LCC lost its hospitals. We were henceforth engaged more in collecting vital statistics, particularly the statistics of infectious diseases, and in the fields of maternity and child welfare and the school medical service. But immediately before this change I had helped in the formation of a new professional body for hospital medical records officers, and I became their first chairman. The emergence of this organisation has been fully documented by Elsie Royle-Mansell, the first secretary [14]. My motive was to improve the level of service given to the clinician, and thereby indirectly to the care of patients; I also wished ultimately to improve the quality of hospital statistical data.

I began to receive outside recognition. Sir Allen arranged for me to address the Society of Medical Officers of Health on statistics in public health. The Royal Statistical Society invited me to read a paper in 1949 on "Local Government Statistics with Special Reference to the Health Service." I contributed to the Royal Society of Medicine's discussion on the influenza epidemic of 1951. In 1950 I was awarded a Rockefeller Foundation Fellowship to tour the USA and study its public health schools and local health authority statistical services. As an actuary I was invited to carry out the third in a series of authoritative follow-up investigations of tuberculosis treatment at Brompton Sanatorium [13].

5. From the LCC to the Government Statistical Office

While I was gaining the recognition of the medical profession, there were difficulties with the administrators. There was, in fact, no professional post of statistician in the public health department; I was graded as a principal assistant. The first time I published a paper as "Statistician, LCC Public Health Department," I was sent for by the Establishment Officer and reprimanded. I protested in vain that I had been debarred from consideration for promotion to a higher *administrative* post because I was regarded not as an administrator but as a statistician.

Reluctantly, in 1952 I sought employment elsewhere, and applied for a post in the recently emerged Government Statistical Service, then under the direction of Harry (now Sir Harry) Campion. He allocated me to fill a vacant post of medical statistician at the General Register Office (GRO). This was normally occupied by a medical practitioner with statistical expertise, but the GRO had been unable to find such a person. The Chief Medical Statistician at the time was Dr. W.D.P. Logan who subsequently became head of the statistics division of WHO. We formed a close working relationship and have maintained our personal friendship to this day. Together we wrote two papers which received wide attention, one on poliomyelitis [16] (then still a problem) and one on a new type of mortality index which was officially adopted and is still regularly published [15].

Though I was, at this time, unique in occupying a post normally filled by a medical practitioner, there was, outside the General Register Office, and to some extent within it, a substantial degree of non-acceptance by the medical profession of the value of the lay statistician. There was still the belief that a layman could not understand enough medicine to be allowed to do sums about diseases with any safety. In self-defence I enrolled at the London School of Hygiene, under the supervision of Professor Bradford Hill (later Sir Austin Bradford Hill), for a Ph.D. in medical statistics. I prepared a thesis on the control of tuberculosis, a disease which I had studied in detail with the help of a number of my medical friends, and gained my higher degree in 1954.

'Tony' Bradford Hill had been a friend and mentor for many years, and still is. The time I spent studying with him was a period of great pleasure, and I learned from him many things apart from the vigorous pursuit of objective analysis. I learned professional humility and discretion; how to build on mistakes (and even good statisticians make mistakes) and not to run away from them. I learned something about honesty. I was only one of his many fortunate students. About this time too, I had joined 'Tony' Bradford Hill and Ronald George as an Honorary Secretary of the Royal Statistical Society. I learned something else too—dedication and loyalty to a learned society for the advancement of a field of learning.

But the Ph.D. soon became irrelevant; being an actuary suddenly became more important. In 1954 V. P. A. Derrick, who was an actuary and the

Chief Population Statistician, retired. To my very real surprise I was promoted to the post he had vacated. For a time I ceased to work in medicine and became absorbed in population problems, though I continued to interest myself in health statistics. Allen and Unwin had invited me to rewrite Newsholme's *Vital Statistics*, then long out of print. I worked on this in my spare time. Also in my spare time I collaborated with the late Arnold Sorsby, Research Professor of Ophthalmology at the Royal College of Surgeons, in producing a series of papers on visual function. He became a very dear friend who is greatly missed.

About this time too, a number of statisticians working in the medical field, some medically qualified, some not, gathered in an informal discussion group to talk about statistical problems peculiar to medical studies. Since most of us were Fellows of the Royal Statistical Society, I approached Professor Bradford Hill (as he then was), the senior Honorary Secretary of the Society, to explore the possibility of forming a Medical Section of the Society. He replied that if we could be sure that the enthusiasm would last and be reflected in good attendance figures, we could go ahead. We were sure and so formed the new Medical Section of which I became the first chairman. We met monthly in the Westminster Hospital Medical School and drank beer afterwards at the Paviours Arms. Those of us working in the medical field no longer felt isolated; the exchange of experience helped enormously to improve both our competence and our confidence.

6. Some Problems in the GRO

My population work from 1954 to 1963 is not within the main theme of this paper, so I do not propose to go into it in any great detail. But it is necessary to give a brief account of the severely restrictive working conditions which were imposed by successive Registrars General from 1946 until the appointment of Michael Reed in 1963.

When the Statistician Class was first introduced in the Civil Service in 1946 its members were specifically regarded as coming within the Administrative Class, and in the statistical divisions of most government departments they administered their own work and directly controlled their supporting clerical staff. The then Registrar-General took a contrary view; that they were professionals who should not be allowed to administer but only to act as statistical advisers to career administrative staff. I was not, therefore, in charge of my own workshop. My main grade statisticians and I were asked only for advice; if it was not accepted there was little we could do. In theory we could not *order* a calculation to be done; we could only *ask* an administrator for its execution. In practice we overcame the difficulty by friendly co-operation.

Communication with statisticians of other departments was often taken over by administrators. If we were allowed to attend an inter-departmental meeting we were usually "shadowed" by an administrator. Often the administrator attended alone. We were being kept away from the coal-face of decision-making. It was irksome and not conducive to good professional work; it inhibited statistical innovation.

The peculiar working conditions seriously affected the execution of the 1961 population census. At that time most computer programming was laboriously conducted in machine language, and it was estimated that the whole programming for the census was likely to take four years. On the basis of a guess at the questions that were likely to be approved, I drew up a tentative schedule of tabulations that could have been used to commence programming (with adjustments to be made later when the census regulations were approved). The administrators of the day were, however, unwilling to approve this schedule until the regulations were actually passed by Parliament in 1960; in fact, programming did not begin until just before census day.

This made it certain that tabulations would not be completed until 1965. The data-processing staff made heroic efforts to catch up; some work was contracted out to a commercial processing firm, and staff were also trained in COBOL to speed up the programming. But indecision had produced irreparable damage. It is generally believed that the delay in producing the 1961 population census results was almost wholly due to the fact that the GRO had to share a computer with the War Office, which had underestimated the growth of its work and had taken more than its original estimated share of running time. Since this work was concerned with the payment of wages and salaries, it necessarily took priority. This undoubtedly caused great difficulty; but if we had been ready to start earlier there would have been fewer problems from this source.

I must make it clear that, as persons, the administrative staff from the Registrar-General down were likeable, agreeable and intellectually talented people whose friendship I valued then, and still value now. The fact that I often wrote them minutes of considerable acerbity in no way diminished my friendly feelings. It was the system that was wrong, not the individuals. I challenged the system repeatedly but without avail. Eventually in 1963 a post as head of the statistical division of the Ministry of Health became vacant. I applied for it and was appointed. Ironically, just at that time Michael Reed arrived as the new Registrar-General and immediately gave me authority to direct my own workshop. But it was too late. I had already agreed to return to my old love—the health service.

7. The Statistical Division, Ministry of Health

This was a difficult but exciting assignment. I had long held strong views on the necessity for comprehensive statistical monitoring of the effectiveness of the National Health Service, which consumed large financial resources. The need for monitoring was, in my view, reinforced by the fact that the resources though large were never adequate; choices had to be made and these choices could hardly be rational without relevant information. I had reviewed the possible sources of data and the ways in which these data could be processed to information in my book *Vital and Health Statistics* [4], first published in 1959 (i.e. four years before I took up this position) and I yearned for the opportunity to build and operate the necessary statistical system.

The statistical division of the then Ministry of Health was a "late developer." From the inception of the National Health Service (NHS) in 1948 until 1955, there was no formal statistical division. The situation remained as it had always been; each administrative division improvised statistics as and when they were thought to be necessary. If you were at the receiving end of a statistical return sent out from the Ministry of Health (as I had been at the LCC) it was very difficult to track down the administrator who had called for the return. It therefore proved difficult to resolve any doubts about the meaning of the questions, and as these were not designed by professionals there often were doubts. There was a high probability of error and little attempt to check for the existence of error.

In 1955 (from memory) the Treasury indicated that this state of affairs was not good enough; it insisted that the Ministry of Health should have a formal statistical division, exactly as other departments had since 1946. As a result, in 1955, John Wrigley was appointed as head of a formal statistical division with a small group of professional statisticians and a supporting staff of clerical and executive grades. The staff allocation was not adequate for proper coverage of the NHS, and an excessive proportion of the limited resources was absorbed by the collection of routine statistics. These were appropriate only to the presentation of global figures of work done for the Annual Report of the Ministry of Health, and had nothing to do with the development of a rational approach to the problem of priorities in the NHS.

But John Wrigley, with the help of Mike Heasman, a medically qualified statistician who held a research development post and sat in the statistical division, had begun to innovate in the direction of producing statistics more appropriate to decision making. In particular, they had instituted a pilot study in a particular hospital of a system of rapid feedback of discharge analyses which formed the basis of what is now known as Hospital Activity Analysis [11].

I set about the task of directing the work of the statistical division on the path of assisting decision making. I made a point of consulting my administrative colleagues on ways in which it might be possible for the statistical division to be brought closer to the policy-making area, without preempting the administrator's responsibility to make actual policy proposals. I was much encouraged by the then Permanent Secretary, Sir Bruce Fraser, who made me a member of his Management Board, consisting of himself, his deputy, the Chief Medical Officer, his deputy, and the other seven under-secretaries. This Board met every Monday evening and discussed new policy issues.

Before any action was instituted, Sir Bruce would ask the appropriate members of the Board to consider what information they needed. He would then ask me (a) if, on the basis of my professional experience and knowledge of available data, I had any suggestions to make as to the appropriateness of the information demanded, and (b) whether I could undertake to deliver the information. Though it meant making bricks without straw, I always said "yes" to (b); and always, the next morning, Sir Bruce would come through on the telephone with the question "Benjamin, how much harm have I done to your work programme—what can we do to help?"

He could not provide extra professional staff, but he could and often did, rustle up a few extra clerical staff to help with the all-important business of collecting and processing data. Most decision-making statistical work has nothing to do with probability or mathematics; it is mostly a matter of getting one's hands dirty in the search for good basic data and cleaning the data up. The stage of analysis is often quite elementary and makes little demand on statistical theory.

I encouraged the appointment of a qualified statistician in every Hospital Region so as to improve my contact with the people in the field. I began to hold regular meetings with them, especially to discuss the development of Hospital Activity Analysis. I knew that I had a difficult road to travel and that I needed support from workers at the hospital level. For there was a vicious circle to be broken. To put a management-oriented statistical system into efficient operation, the consultants must be made not merely to tolerate the system but to become convinced that it would really help them to improve the management of their departments and their patient care. On the basis of this conviction they could then inspire all grades of staff below them to co-operate.

The consultants, many of whom were nearing the end of their careers were, however, less willing to innovate than their juniors. These juniors, in their turn, were not prepared to innovate, and certainly not to accept the discipline involved (e.g. the prompt completion of case-records) without strong pressure from above. I stumped the country talking to hospital consultants about the use of statistics in management (*their* management), but their initial reaction was poor. It was common for them to refer to "your" statistics and I had to stress repeatedly that the feedback policy was based on the belief that the statistics were "theirs," not mine.

Gradually we gained sympathy and even some enthusiasm. The regional statisticians offered enormous support; they became our peripatetic "trouble-shooters" who could tour hospitals to advise on difficulties encountered in the completion of the discharge summaries and in the coding, especially of diagnoses. This too had the indirect effect of convincing the junior medical staff that their work in the completion of records was not unnoticed and indeed was subject to certain "expectations" at the regional level.

There is today, a wide coverage of Hospital Activity Analysis and an increased awareness among medical practitioners of the need for statistical selection of the appropriate strategies for both administrative and clinical management. But there are new difficulties. Economic pressures have made medicine, in common with other professions, less vocational and more acquisitive. We are all in danger of becoming uninspired "nine-to-fivers." In addition, the callous limitation of the share of national resources allocated to the National Health Service has so crippled it that morale has fallen below the level at which good managerial practices are fostered. Efficiency, especially medical efficiency, cannot be afforded. Few feel encouraged to try. However, what has been demonstrated, cannot be retracted; it will receive recognition and eventually become popular again.

I encountered other difficulties. It seemed to me to be very important to do something about the statistics of primary care, i.e. of general practice both in medicine and dentistry. As Director of Statistics I was retained as the Ministry's professional adviser in the regular pay reviews of both doctors and dentists. I collected a great deal of information about their respective work loads, and my evidence about work loads and expenses was always accepted by both bodies.

In those days Sir Roy Allen was the independent statistical adviser to both bodies. I was always able to reach agreement about my evidence with him in advance. This did not avoid argument about remuneration, but the prior agreement of the two professional statisticians about the basic facts did much to reduce it. But in relation to the monitoring of the Health Service, there was a serious difficulty connected with general medical practice. We knew that general practitioners were busy, but we did not know what their business involved and to what objectives it was directed [3]. It was not even possible to persuade general practitioners to put a diagnosis on their prescription forms so that we could at least know for what therapeutic purpose a drug was being administered. And we had bad luck.

At one time Sir Bruce Fraser set up a working party on general practice and as a member of this group I persuaded the representatives of the general practitioners to accept a proposal that a sample of practices should be "observed" by acceptably trained persons (retired doctors or nurses). We even agreed on a questionnaire that would be completed in respect of each patient consultation. But almost immediately discontent over doctors' pay flared into open warfare; co-operation was withdrawn and the enquiry aborted. Subsequently the Royal College of General Practitioners was established, and a number of studies of general practitioner work were encouraged. However, the most representative of these have been concerned only with disease prevalence and not with the direction or effectiveness of primary care.

The third branch of the National Health Service—community health care, as it is now called—had been in the doldrums since 1948 and by and large still is. There was a pattern of local social and medical services from antenatal and paediatric care to old peoples' homes, some of them statutory and some voluntary, and also a pattern of local population needs. No one knew, nor do they yet know, how well or badly these two patterns meshed together. Local authorities were not prepared to spend money on studying the extent of the fit. The voluntary bodies, at least those that I consulted, were also not willing to do so.

There were further complications. No data-processing facilities existed which were dedicated to the statistical division and under my control. The Department of Health was in the process of commissioning a computer but this was to be a common service unit, an anathema to any Director of Statistics worth his salt. For the duration of my stay at the Department, nine different commercial computer agencies were used. Since our work was highly technical, this meant that we had nine separate problems of communication. One agency, for example, working for most of the time on commercial costing or market research statistics, handled our national index of psychiatric patients with a highly technical range of diagnostic labels. On the whole, the agencies were remarkably competent and helpful though their endeavours to spread overheads led them to overload their facilities and so to break deadlines.

All the time I was crystallising my own ideas on monitoring health services; and my own staff were learning what it was that I was trying to do. I failed, however, in at least one serious respect. While I personally continued to fight the idea, still sedulously followed by many of the doctors within and outside the Department, that only a medical practitioner could be trusted to make statistical calculations about disease prevalence and disease treatment, I was often accused by my medical colleagues of being paranoid about my conviction. Indeed, I was hypersensitive and often, to my present shame, intemperate; but it was a very important tenet of my belief in the advancement of statistics as a subject that its principles did *not* vary with the field of application. To convince doctors that lay statisticians were perfectly competent to handle their medical statistics, I would listen to medical practitioners and try to demonstrate that I understood their problems. But I still could not persuade my professional subordinates to listen and learn in the same way. They could not spare the time, that is, they could not see the importance of organising themselves into sparing the time.

There was another problem not confined to the Department of Health, though it was more acute there than in other Departments as an aftermath of the statistical extemporisation from 1948 to 1955. This was the sporadic eruption of statistical activity outside the statistics division. There were still occasions when an administrator would decide to conduct his own statistical enquiry in some sector of the Health Service. It is true that there were faults on our side. Preoccupied as they were with the collection of routine statistics often unrelated to policy making, my statisticians had not completed a sufficient number of really useful jobs to impress the administrators with their capabilities. Regrettably, when approached by administrators, they had often turned them away on the pretext of being too busy. As a consequence, we acquired the image of being remote (we were in a building a bus ride away from headquarters) and uncooperative.

For their part, administrators did not understand that even the design and processing of a simple questionnaire was likely to be more efficiently, accurately and economically handled by a skilled statistician. They rarely appreciated that a statistician has been trained to ask questions, to validate replies, to arrange efficiently the apparently simple task of combining figures from individual returns. On one occasion a working party, without consulting me, committed clerks in hundreds of hospitals to completing questionnaires which could not provide the information that the enquiry was set up to discover. They had to start all over again, this time with our advice.

I cajoled my staff into never refusing a task without first consulting me. If it was a small task, we should try to fit it in somehow. What I call "quickies" are a very important part of good public relations; they win you friends with minimal effort. Unfortunately they also win you more customers who further overload your limited resources. If it was a large task we should not refuse without first trying to obtain additional resources; usually the resources were found somehow. In my experience (and possibly I was spoiled in this respect) the British Civil Service is extraordinarily good at producing emergency clerical help on quite a large scale at very short notice. The problem of "extramural" statistical activity gradually diminished.

I must mention one respect in which I think the Department of Health failed me, though this was possibly due to my inability to make my position clear. Computers were beginning to come into medical practice itself, not only in simple ways like the VDU representation of case-histories in the consulting room, but also in more complicated ways such as the analysis of electrocardiograph or electroencephalograph wave forms. A small party consisting of Dr. Henry Yellowlees (now Sir Henry Yellowlees, the Chief Medical Officer at the Department of Health and Social Services), Dr. Percy Cliff (of the Clinical Measurement Department at Westminster Hospital), Dr. Austin Heady, an epidemiologist (then at the London School of Hygiene and now at the Royal Free Hospital) and myself was given the exciting task of touring medical schools in the USA to see what was being done there, and to assess the potential value of the computer to medical care.

In our report (which I wrote), we suggested the establishment of a computer laboratory in which doctors could explore for themselves the possible uses of the computer and become trained in handling computer equipment. Above all it was urged that a senior officer should be appointed to operate a "think tank" to explore ideas; he should not be a conventional management services type, accustomed to thinking only of batch processing or records storage and retrieval; he should be an imaginative scientist. We did not get our laboratory, nor did we get our "think tank." As a consequence, some computers were introduced in hospitals for wrong purposes or with insufficient prior acceptance by clinicians, and were either underused or abandoned.

During my period at the Ministry of Health I continued to interest myself in medical research and gave voluntary spare-time assistance to friends in the medical profession. It is interesting to note that although some diagnostic centres have now appointed statisticians, it is still difficult for the individual clinician interested in research to obtain statistical help except on this basis. My work with medical friends helped them, but it also helped me to extend my still very limited knowledge of medicine. I particularly valued my collaboration with Professor Lynne Reid and her assistants in the Experimental Pathology Department at the Cardiothoracic Unit at Brompton and with Arnold Sorsby, the Research Professor in Ophthalmology at the Royal College of Surgeons. My work on refraction with Sorsby began in 1948 and we wrote our last paper together a month or two before he died in 1980. In my spare time, I prepared for UNESCO a bibliography on the subject of social and economic factors in mortality, a field only just beginning to be seriously explored at the international level. This was later published as a methodological and reference text [2].

8. Director of Intelligence, Greater London Council

In 1965, my full-time work in public health came to an end. As an old local government man, I had long cherished the hope that statistics would come into its own in local government and that local authorities would some time see the virtue of an independent fact-finding department. The Royal Commission on London Government of 1960 which recommended the setting up of the Greater London Council (GLC) had insisted that there should be an independent "intelligence" department in the new authority. This recommendation was accepted and written into the London Government Act 1963 which created the Greater London Council. So when the Greater London Council advertised the post of Director of Intelligence, I applied for the post and was chosen to fill it. It was a disastrous decision on my part though curiously I do not altogether regret it. The story has nothing to do with the main theme of this paper and so I will tell it only very briefly.

The Greater London Council, which replaced the London County Council and the Middlesex County Council, was to govern a wide region and a population of nearly eight million people. But there were to be two tiers of government. Below the Greater London Council there were to be the 33 London Boroughs (including the City of London) each controlling a population of between 200,000 and 250,000. The London Boroughs were to be the executive authorities, while the Greater London Council was to be a relatively small strategic authority laying down broad lines of policy within which the London Boroughs would operate. That was the theory. In practice, the Greater London Council became a gigantic combination of the staffs of the former Middlesex and London County Councils. It was the London County Council "writ large," gathering up all the executive tasks it could lay its hands on.

Most of the Chief Officers of the GLC were former Middlesex County Council and London County Council men, and were no more appreciative of the need for independent statistical advice than they had previously been. On the contrary, they were resistant to the idea. Whether as a result of the pressure of this resistance, or because the elected members of the Council did not understand the concept, there was a delay of one year between the setting up of the GLC and my appointment. The Chief Officers took advantage of this delay to build up their own in-house statistical units. The Planning and Transportation Department, in particular, had a very large staff engaged on statistical and research work; very few were statistically qualified, most being town planners or engineers. The Establishment Department had an operational research unit. The Inner London Education Authority had its own research unit. By the time I arrived they were well dug in and very resistant to any suggestion that they should be transferred to an independent statistical department.

Moreover my post had been quickly downgraded to Deputy Chief Officer level and I was attached to the Director-General who by tradition was primus inter pares and exercised very little authority over the other Chief Officers. The latter had direct access to their committee chairmen who could exercise political muscle. But though I was not concerned about rank, I *was* certainly concerned at the difficulty of achieving the spirit of the London Government Act, which was to organize a central statistical department serving *all* departments (and incidentally the London Boroughs). I decided that what I lacked in power I could make up for in competence. The departments would find that it was ultimately to their advantage to consult us.

I managed to persuade the Council to let me have a small staff of highly qualified and enthusiastic professionals to achieve this objective. For four years, against much opposition, we worked hard to build up an image of objectivity and competence. Towards the end of this period, it began to dawn on our political masters that a lot of money was being spent on uncoordinated research units in different departments without any assurance of cost-effectiveness. It was finally decided that all the units were to be brought under my control, but only at the price of my moving with my staff to the Department of Planning and Transportation (on the pretext that a large part of our work would be for that department).

So there would be a loss of independence, and a departure from that essential feature of good management which specifies that a central statistical unit should be the tool of the General Manager (in this case, the Director General). Moreover I discovered that there were plans to break up the enlarged unit once it was safely in the Planning and Transportation Department. Promises to this effect had been made to dissident individuals, most of whom were not statisticians. I came to the conclusion that enough was enough and, my sixtieth birthday being close, I retired on pension. The unit was promptly broken up after my retirement, the retirement being given as an excuse for this action! Some years later when Sir James Swaffield took over as Director General of the GLC he recovered the surviving rump of my unit and restored it in its proper place—at his elbow.

9. The Civil Service College

I was fortunate in being immediately invited to become the first Director of Statistical Studies at the newly inaugurated Civil Service College with the task of teaching administrators how to use statisticians! I was also invited to become the next President of the Royal Statistical Society. From 1966–68 I had been President of the Institute of Actuaries, so there was extra pleasure in becoming the first actuary to be president of both bodies. All this mitigated the great disappointment of unfulfilled hopes, but I cannot disguise the fact that it *was* a bitter disappointment.

I had, of course, made mistakes. I was not always as tactful or tolerant as I should have been; and I was in too much of a hurry to create an ideal system in the far from ideal world of local government. It is often alleged that the level of management skills in British local government is lower than in the central government. Whether this is true or not, certainly the Chief Officers in the GLC in 1965 did not seem to understand the relationship between policy making and information. They were even unwilling to recognise that the statistician was a professional with training in the interpretation of management needs, and with jealously guarded standards of objectivity, professional competence, and integrity. It was not that they did not desire these qualities. They failed to understand them and were consequently suspicious of them.

I subsequently wrote a booklet which amounted to a statement of what I would like to achieve and the ISI published it [7]. I have also published my credo about the relationship between the manager and the statistician [6]. I was sorry to leave behind the dedicated and loyal staff who had joined me in the early days. Fortunately, I am able to report that most of them moved on to a better statistical environment and in most cases to more senior appointments.

My story is not quite ended. My experience at the Civil Service College was most enjoyable. To have passed from a situation of great tension (my blood-pressure must have been very high) to a quasi-academic environment was a piece of great good fortune. I gave up smoking my pipe, which had been the subject of caricature by a cartoonist on my GLC staff. It was also a wonderful opportunity to put into practice some of the lessons I had learned in the GLC. With the help of a few handpicked lecturers, we taught administrators the language of the statistician without using mathematical terms [5]. The object was to enable them to communicate their needs to the statistician and to understand his response. We taught government statisticians to understand the relationship between management and information. We also inaugurated refresher courses in statistical methods for statisticians who had got out of touch with developing theory.

10. Actuarial Science at City University

Then in 1973, the Life Offices Association with the support of the Association of Consulting Actuaries decided to endow a university chair in actuarial science and I was asked to be the Founder Professor. It had to be done quickly as I had not much time left for full-time work. My wife was in failing health: she died some three years later. We decided therefore to go to the City University, newly created in 1967 out of the former Northampton College of Advanced Technology and independent of the University of London, which was prepared to start actuarial science teaching immediately.

With the generous help of the Professors of Economics (Colin Harbury) and Mathematics (Maurice Jaswon) who subsequently provided service courses, I quickly devised a new degree course leading to the B.Sc. in Actuarial Science. The concept was not new: such a course had been started in 1966 at Macquarie University, Sydney, Australia by my friend Professor Alfred Pollard, to whom I am especially grateful for allowing me to build on his experience. The degree course was established in its own right and covered mathematics, statistics, economics and some specialist actuarial subjects. But it also carried with it the possibility of exemption from the Intermediate examinations of the Institute of Actuaries. The university accepted this, and initially with the help of only one other actuary (Steven Haberman who currently directs the actuarial science school) we began teaching in October 1974.

The course has been a great success story, and the school now turns out 30–35 graduates every year, most of whom go on to complete their professional qualification. Other universities in Britain, such as Exeter and Kent, are now in the process of copying us. In 1975 I handed over the full-time reins to Professor James Pegler who had just retired as General Manager of the Clerical, Medical and General Assurance Society. He and I are still visiting professors, and two additional actuarial lecturers have joined Steven Haberman.

I had found time, miraculously, during my sojourn at the GLC to write the official textbook on the analysis of mortality and other actuarial statistics for the Institute of Actuaries [14]. This is the most difficult of all subjects in the Intermediate examinations of the Institute of Actuaries. Previous attempts to produce a satisfactory teaching text had failed. This attempt seemed to succeed. Possibly for this and for other contributions to the actuarial literature, the Institute of Actuaries honoured me with the award of their Gold Medal in 1975. In 1976 I wrote their textbook on general insurance [8], a subject newly introduced in the syllabus. It was a crash job; the subject had to be introduced in 1978 to meet growing professional demands. So I did six months' reading and six months' writing. There were all the ingredients of a disaster, but somehow with the help of my Institute colleagues it turned out not too badly. At least the book has since been reprinted.

11. Concluding Remarks

All this time I have retained my interest in medicine. In 1977 I published, as editor and author of the bulk of the text, a textbook on medical records [11]. I also gave an address to the Royal Statistical Medical Section 100th meeting on "Progress in medical statistics" [9]; reprints of this have been in demand all over the world. I have continued to act on occasions as consultant to WHO in Geneva [10], [12]. I have retired three times, but never will. As actuary and statistician, I have always found my work fun and its cessation unthinkable. Despite my diversions I have always regarded myself essentially as a "public health" man; if I have helped in any way to win the war of the "damned dots" in medicine, I am amply rewarded.

Publications and References

[1] Benjamin, B. (1943) Height and weight measurements of schoolchildren. *J. Hygiene* **43**, 55–61.

[2] Benjamin, B. (1965) *Social and Economic Factors Affecting Mortality*. Mouton, The Hague.

[3] Benjamin, B. (1967) Assessment of medical care. *Proc. R. Soc. Med.* **60**, 809–813.

[4] Benjamin, B. (1968) *Vital and Health Statistics*. Allen and Unwin, London.

[5] Benjamin, B. (1974) The statistician and the manager. *Omega* **2**, 263–267.

[6] Benjamin, B. (1975) Teaching official statistics. From *Teaching Statistics at the School Level*, ed. L. Råde. Almqvist and Wiksell International, Stockholm.

[7] Benjamin, B. (1976) *Statistics and Research in Urban Administration and Development*. ISI, The Hague.

[8] Benjamin, B. (1977) *General Insurance*. Heinemann, London.

[9] Benjamin, B. (1977) Progress in medical statistics. *J. R. Statist. Soc.* A **140**, 366–367.

[10] Benjamin, B. (1977) Trends and differentials in lung cancer mortality. *World Health Statist.* **30**, 118.

[11] Benjamin, B. (1977) *Medical Records*. Heinemann Medical Books, London (Second edition 1980).

[12] Benjamin, B. (1978) Tobacco smoking in the world. WHO Expert Committee on Smoking Control, CVD/8/PC/78.23. WHO, Geneva.

[13] Benjamin, B., Foster Carter A. F., Myers, M., Goddard, D. L. H. and Young, F. H. (1952) The results of collapse and conservative therapy in pulmonary tuberculosis. *Brompton Hospital Reports* **21**, 1–49.

[14] Benjamin, B. and Haycocks, H. W. (1970) *Analysis of Mortality and other Actuarial Statistics*. Cambridge University Press, London.

[15] Benjamin, B. and Logan, W. P. D. (1953) Loss of expected years of life. *Monthly Bull. Min. of Health* **12**, 244–246.

[16] Benjamin, B. and Logan, W. P. D. (1953) Geographical and social variation in poliomyelitis. *Brit. J. Prev. Soc. Med.* **7**, 131–136.

[17] Benjamin, B. and Pollard, J. H. (1976) *Analysis of Mortality and other Actuarial Statistics*. Heinemann, London.

[18] Daley, Sir Allen and Benjamin, B. (1942) Tuberculosis in London in wartime. *Brit. Med. J.* **2**, 417–420.

[19] Gore, A. T. (1966) Seventy-six years of public health in London. *Medical Officer* **CXVI**, 105–114.

[20] Spear, B. E. and Gould, C. A. (1937) Mechanical tabulation of hospital records. *Proc. R. Soc. Med.* **30**, 633–640.

Henry Oliver Lancaster

Professor Henry Oliver Lancaster was born on 1 February 1913 in Sydney. After completing his secondary education at West Kempsey High School, he entered the University of Sydney and started on an Arts course. At the end of the year, having obtained a scholarship, he opted for medical training, which he completed in 1936 when he graduated MB, BS.

He started his career as a resident medical officer at Sydney Hospital in 1937, and remained there as a pathologist until 1939. It was then that he began to take an interest in statistics. In 1940 he joined the Australian Imperial Force as Medical Officer and served in the Middle East and New Guinea until 1946.

At the beginning of 1945 he put himself through a serious course of study in mathematics at the University of Sydney. In 1946, he was appointed Lecturer in Medical Statistics at the School of Public Health, University of Sydney, and over the next decade rose to be Associate Professor. In 1959 he was appointed to the first Chair of Mathematical Statistics in this University; he occupied it until his retirement at the end of 1978.

Professor Lancaster has successfully combined his medical and mathematical interests in his statistical work. He has played an important role in Australian mathematics and statistics: he was the founding editor of the *Australian Journal of Statistics*, and served as President of both the Statistical Society of Australia in 1965–66, and the Australian Mathematical Society in 1966–67. He is a member of the International Statistical Institute, and a Fellow of both the Institute of Mathematical Statistics and the American Statistical Association. He was elected Fellow of the Australian Academy of Science in 1961 and awarded its Lyle Medal for mathematics and physics in 1962. In 1980 he was presented with the Pitman Medal of the Statistical Society of Australia. He has written numerous papers on mortality, mathematics and statistics and is the author of several books.

He has five sons; his recreation is bowling.

From Medicine, through Medical to Mathematical Statistics: Some Autobiographical Notes

Henry Oliver Lancaster

1. Early History and Education

I was born in Sydney, New South Wales, Australia on 1 February 1913, the second son of Llewellyn Bentley Lancaster and Edith Hulda Smith. My maternal grandfather Henry Jorgensen (later Smith) was born in Moss just south along the coast from Christiania (Oslo) in Norway. He had been an officer in a vessel trading to South Australia and later a leading manager of Millars, a timber-getting firm in the Albany and Bunbury areas of Western Australia. My maternal grandmother was Matilda Hülda Mattner; she belonged to a group of migrant families who had resisted the changes in liturgy in the Lutheran church and also had had conscientious objections to military conscription; her ancestors came from an area roughly midway between Frankfurt on the Oder and Poznan (Posen) now in Poland. The story of some of these families has been recorded in *The history and family tree of Johann Wilhelm Rohrlach and his wife Anna Dorothea née Galpach and their descendants, 1792–1980* [32] and a similar book in the press on the Mattner family.

My paternal grandfather James Lancaster had come as a boy migrating with his parents to Port Macquarie on the coast of New South Wales about 300 km north of Sydney and then to the Macleay River a few miles further north. My paternal grandmother Jane Norton had also migrated to Australia as a child. My father was the youngest of their six children; his father died when he was six years old. However, he was able to complete a good education first at Newington College in Sydney and then at Sydney University. He graduated the youngest in his year MB, Ch.M. After a residency at Sydney Hospital and practice in Western Australia and Queensland, he returned to Kempsey on the Macleay River in 1904 to practise. The depression of 1893 had caused severe economic stress in the area, but the period from about 1900 until 1930—and so until his death in 1921—were the golden days of country medical practice and practitioners.

In the isolated areas, the society of the towns was dominated as a rule by the trained professional men, who had been educated in the cities. By the time of my birth, it was practical to motor down the coast to the railway system centred on Newcastle and then cross the Hawkesbury River by punt and continue by train to Sydney. My parents did so and my father played a challenge match against the Australian chess champion, W. S. Viner, losing by seven games to two, a result surpassed by no other challenger.

I was thus born in Sydney into a prosperous family, well known and respected on the Macleay River and surrounding areas. I and my two brothers, Geoffrey and Richard, respectively two years older and one year younger than I, attended a private school conducted by the daughter of the Presbyterian minister and each of us developed rapidly. My facility in mental arithmetic soon became apparent and I was reading fluently by the age of six. The skill at simple sums was a source of amusement to my father's professional colleagues from Sydney who visited our home in Kempsey on their way to the fishing at the mouth of the Macleay River. At an early age, I developed a stammer which people have not noticed much in my later life but which was a great source of embarrassment in my earlier life.

My father died just before I attained my ninth birthday. After his estate had been settled, my mother decided that she would be forced to complete her double certificate in nursing, so Richard and I stayed at a boys' hostel in Kempsey, where we greatly enjoyed our school-days; we were well cared for and played a good deal of cricket, tennis and football. The education at West Kempsey Intermediate High School was good, for many of our teachers had good arts or science degrees.

This era was good for the country high schools although they did not serve a wide enough section of the community. Indeed, there were only six students reaching matriculation from a potential child population about the same size as that now in the same area, which now provides over a hundred matriculants annually. Our final year class had no permanent classroom but used to occupy the room left vacant by the class doing science in the special science room, or an open verandah looking down onto the playground.

In the leaving (matriculation) examination for the State of New South Wales, I was *proxime accessit* in mathematics to John M. Somerville, later professor of physics in the University of New England, and third in chemistry, and so rated as having received three first-class honours, together with A's in French, English and Mechanics and a B in Latin. Excepting this last, this was the maximum allowed under the rules. I received a bursary of £65 per year and a one-year scholarship of £50 and exemption from fees. These amounts would have seen me safely through first year at the University of Sydney if I did not attend one of the colleges, which in those days had little sympathy for those unable to afford regular fees.

2. A Student at the University of Sydney

I wished to study science, although there was a sentimental leaning to medicine, but could only count on full support for one year. My mother, advised by men truly eminent in their respective traditions, wished me to do an actuarial course. I therefore enrolled as an evening student in economics, being prepared to drop my bursary and scholarship to do so, and took a position in a leading life assurance office. After three weeks, I decided that I could do at least one year as an arts student and return to the assurance office at the end of the year. I therefore began an arts course in the fourth week of first term with Mathematics, Chemistry, Economics and English as my subjects.

At the end of the year, I won the Struth Exhibition of £50 per annum and began a medical course. This imposed a financial burden on my mother because my exhibitions and scholarships were not sufficient to support me through the course. With hindsight, one sees now that the science course should have been chosen initially, come what might. However, the depression of 1930 might be recalled; and my older brother in Western Australia had had similar problems after an equally good performance at matriculation level. It may also be recalled that the outlook for academic mathematics was not at that time promising in Australia. In any case, I was glad and even anxious to do medicine in the hope that it would perhaps lead to an honorable professional life and a proper reward.

In general, I liked the medical course, but second-year studies were unduly dominated by anatomy; I do not have a particularly good visual memory nor do I sketch well. I disliked the detail that had to be memorised and, in particular, the fortnightly viva voce examinations in which I felt I was at a disadvantage because of my stutter. In third year, the professor of anatomy was A. N. St. G. Burkitt, an excellent scholar and teacher, who was fond of recalling the recent eminence of the department under J. T. Wilson and J. I. Hunter and who introduced us to the comparative and evolutionary aspects of anatomy.

Many years later I was unsuccessful in convincing the university to celebrate the seventieth anniversary of Hunter's birth more than forty years after his untimely death. But I was successful in persuading the university to celebrate the centenary of Grafton Elliot Smith, to which several eminent overseas anatomists were invited. That this was reasonable was shown by the parallel symposium held in London under the chairmanship of Professor Solly Zuckermann. The comparative studies of Smith on the marsupials had led him on to far-reaching generalizations in evolutionary theory and anatomy. He overstated the evident case for the diffusion of culture and in particular erred in assigning a central position to Egypt in the evolution of ideas. There was adequate evidence at the two symposia of the value of his work, especially in anatomy and medical education, in both of which he had been a leading figure.

I found physiology interesting, but the great days when important work could be done with very little equipment or knowledge in the other sciences were ending; the department was in decline. Biochemistry was too divorced from organic chemistry and tests for the presence of particular chemicals in the body therefore seemed to be pulled out of the hat. I probably read too widely in the second and third years; the classics of Pavlov and Sherrington, although recommended for additional reading, in no way helped in the immediate future. I should have liked to do a special B.Sc. course in a physiological or biochemical topic but my limited finances would not permit this.

3. Clinical Studies and Residency in Medicine

In the fourth year, we began our clinical studies. Bacteriology, pathology, surgery and medicine lectures were given in the morning and then there were visits to the wards and specialized clinical lectures in the afternoons. I especially enjoyed bacteriology and read *An Introduction to Immunity* by W. W. C. Topley, an extended version of parts of the well-known Topley and Wilson *Principles of Bacteriology and Immunology*. I could see that there was scope for chemical and physical applications there and hoped later to do research in that field. Our clinical education continued in the fifth and sixth years.

The professor in obstetrics, J. C. Windeyer, did much to popularise modern, especially preventive, methods. Many years later, a grandson, John Windeyer Donovan, secured a Ph.D. under my actual, but not nominal, direction for a historical study of mortality in New Zealand; under the rules of the National Health and Medical Research Council, he could not study nominally in a non-medical department.

In the fifth year, the Professor of Preventive Medicine, an intelligent and widely read man with conservative but humane ideas, was Harvey Sutton, greatly respected in the community at large and known by sight to many cricket fans and others of the 1930s. I was to think in later life that he was gravely handicapped by his honesty or lack of guile, which prevented him from practising the administrative arts skillfully or even suspecting dishonesty in published work.

Indeed, if I were asked to nominate my preferred lecturers, I would nominate T. G. B. Osborn in botany, A. N. Burkitt in anatomy, H. D. Wright and N. E. Goldsworthy in bacteriology, C. G. Lambie in medicine and Harvey Sutton in preventive medicine. By my fourth year, I had regained the confidence I had lost in the whirl of detailed anatomy in second year, and rather enjoyed the clinical work of history taking and diagnosis; and I was rather proud of my memory for the faces of former patients.

I had been a student at Sydney Hospital, founded 1811, and elected to be a resident medical officer there in 1937 since my father had been such in 1896; also I was closer to my home in Manly. This was possibly a mistake although several physicians and the group in clinical pathology were excellent. In my first year as resident medical officer, I entered for the New

South Wales chess championship winning about four games out of a possible eleven.

One victory was against a young graduate in chemistry, J. W. Cornforth, later a Nobel Laureate. In 1961, I met the radio astronomer B. Y. Mills who professed to remember me well, since as a schoolboy he had been an interested spectator of the game just mentioned. In 1977, when Cornforth was being entertained by the New South Wales Group of the Fellows of the Australian Academy of Science, I was able to have a photo taken of the three chess players forty years after. However, a chess tournament combined with a heavy junior resident post proved far too strenuous and I have not played in competition since.

I was resident pathologist at Sydney Hospital in 1938 and 1939 and was very happy with the life. Our director was J. C. Eccles (later to be a Nobel Laureate) who had been appointed Director of the Kanematsu Institute at Sydney Hospital and, during the years of his directorship, the Institute had a very strong research record. Among others in Eccles's staff there were B. Katz, S. W. Kuffler and J. O'Connor, who all attained eminence in neurophysiology. O'Connor died young but Katz became a Nobel Laureate and also Vice-President of the Royal Society, while Kuffler became a foreign member of the Royal Society and member of the U.S. National Academy of Sciences.

In 1939, I attempted to combine my duties with a university course in second-year chemistry but the practical difficulties were too great. However, I read Udny Yule's *Introduction to the Theory of Statistics* with some understanding, but the appearance of $\surd(2\pi)$ suggested to me that I would need to resume some part of my mathematical studies.

In 1940 I went as a Junior Fellow in Medicine of the Postgraduate Committee in Medicine to the Prince Henry Hospital, Little Bay, under a novel scheme to allow the Fellow to pursue some theme in clinical medicine, which for me was the clinical background to pathology, and to take the MRACP examination (for physicians). War pressures prevented the proper development of the scheme and my participation in it. I went as resident pathologist to Royal North Shore Hospital, where in my spare time I considered the rapid sedimentation of the red blood cells. Others finally solved the problem with the same principle but with the use of a separating machine as for milk.

4. War Service

On 31 July 1940 I joined the Australian Imperial Force as Medical Officer and served until April 1946; I spent one year in the Middle East and two years in New Guinea.

At first I was medical officer to a general hospital but it was recognized that there was or would shortly be a shortage of pathologists and so I was sent to an army school in pathology and went abroad as pathologist to the 9th Australian General Hospital. Of course, conditions were rough for pathology under war conditions as in the "khamseen," the hot dusty desert wind, outside Alexandria, but the required laboratory work could nevertheless be carried out with appropriate care. Our unit was to go to Greece but our convoy did not leave Egypt. We were greatly impressed by the visit of Brigadier J. S. K. Boyd, later FRS, with several senior colleagues to our unit. At this camp as elsewhere, I helped to organise clinical or pathological meetings to help maintain medical interest.

One of our unit, Major (now Sir Douglas) Miller did distinguished neurosurgery on the western front, it being found necessary for the surgeon to be well forward in the fighting area. Some members were exchanged for their opposites in the 4th Australian General Hospital at Tobruk during the siege in 1941. However, their pathologist was working on relapsing fever and did not wish to be relieved. Our hospital was next established at Nazareth.

In February 1942 we returned to Australia. At the 117 AGH in Townsville, Major T. E. Lowe and I investigated the incidence of eosinophilia and found that it was usually caused by hookworms or strongyloides. I also investigated the incidence of intestinal protozoal infections. Here a statistical problem was to compare incidences in the four classes of troops according to whether or not they had been to the Middle East or New Guinea, a problem in $2 \times 2 \times 2$ tables; higher orders could occur if several parasites were considered.

After a relieving stretch I went to New Guinea in 1944. Here my first task was to assist in the determination of the cause of an epidemic of certain fevers in the Finschhafen area. I was greatly impressed by the epidemiological methods of the clinician Major H. Love, who showed that some combination of certain symptoms characterizing sandfly fever was present in each case; my part was to show that these cases were not undiagnosed malaria. Malaria, scrub typhus and the dysenteries were other diseases occurring during the campaign in the area.

In August 1944, I was seconded to the Australian New Guinea Administrative Unit as pathologist, attached to headquarters at hospitals in Port Moresby and Lae. Contrary to expectations, the diseases in the native troops and civilians were not the exotic diseases thought of as "tropical diseases" but aberrant forms of diseases much studied in temperate zones, for example, glandular tuberculosis, cerebrospinal meningitis and lobar pneumonia. There were also cases of dysentery, malnutrition, anaemias and so on.

When we moved to Lae we were gratified to have weekly visits from a leading surgeon and pathologist, Major E. S. J. King, among others. I made a survey of more than a thousand native troops and civil workers from different areas and surprised the army Director of Pathology, Colonel E. V. Keogh, by reporting the results in systematic form with properly drawn graphs and means and standard deviations correctly computed. Incidentally, the errors made by our visitors in diagnosing the cause of swollen glands in the neck brought home to me that medical diagnosis uses or should use past experience with a change in a priori odds as one moves into a new area; in modern language, sound diagnosis uses Bayesian methods.

5. Mathematical Studies

At the beginning of 1945, I began a serious study of pure mathematics after a break of 14 years. In 1944 I borrowed Caunt's *Infinitesimal Calculus* from our adjutant, C. J. Stevens, now principal actuary of the Australian Mutual Provident Society, and found that I could remember an encouraging amount of mathematics after only a little practice. So, in 1945, I enrolled as external student in the second-year honours course at Sydney University. There was plenty of time in the evening to study mathematics, sitting with a kerosene lamp in hot humid conditions.

My aim was to prepare myself to do a course in chemistry or physics after demobilisation. However, as I had learnt by some pilot investigations, there was much to be done in demography in New Guinea and elsewhere for anyone with a formal medical education. I obtained a high distinction in Mathematics II and later enquiry showed that I had answered as many questions as others obtaining the same result. My ideas on mathematical education were no doubt old-fashioned: I merely read the textbooks and worked out as many examples as I could. Books I remember having liked are Hardy's *Pure Mathematics*, Bromwich's *Infinite Series*, Woods's *Calculus* and Sommerville's *Conic Sections*.

On demobilisation, I followed this up with a distinction in Mathematics III in 1946. In 1947, I attended the pure half of the Mathematics IV course. I fear that my motives in pursuing mathematics so far were suspect in the eyes of some medical colleagues, who feared that I was abandoning medicine for mathematics. In the third-year course in 1946, I attended the courses in applied mathematics under Professor K. E. Bullen FRS, who gave very clear lectures and who was a help and encouragement to me then and for many years later.

I also attended the lectures of Professor T. G. Room FRS in pure mathematics, principally geometry. An older student appreciates some mention of the foundations of mathematics, and geometry is perhaps one of the best introductions to the subject. In any case, I have always been greatly impressed by Room's lectures and his virtuosity. No other lecturer in Australia has seemed to me to have such command of the subject in hand; and among those lecturers heard abroad only a few others, including Professor G. E. H. Reuter in England, have given me the same impression.

At this stage, I certainly had no intention of becoming a professional mathematician. I did think, however, that many well-known applied statisticians attacked problems without the necessary mathematical knowledge and manipulative skill. Moreover, I believed that a principal cause of failure among medical research scientists was the lack of basic scientific knowledge in their special chosen field, and I had no desire to follow such a lead. In the elementary texts, which I was reading at that time, there were numerous examples of such a lack of basic knowledge in the form of clumsy statements, false inferences, errors in distribution and so on. I found M. G. Kendall's first volume [3] far more helpful than the elementary texts.

6. The Sydney School of Public Health and Tropical Medicine

On 1 April 1946 I became a temporary member of the staff of the School of Public Health and Tropical Medicine under the Director, Professor Harvey Sutton, it being understood that my first year would be spent in acquiring expertise in medical statistics. This was interpreted by me, at any rate, as attending Mathematics III and reading the works of the English statisticians such as Farr, Greenwood and Hill and the American statisticians and epidemiologists such as L. I. Dublin, A. J. Lotka, R. Pearl and W. H. Frost.

I thought it would be quite wrong to teach applications of statistics to epidemiology using English or other non-Australian data and so set about systematizing the official Australian mortality series, which had not been comprehensively studied, although G. H. Knibbs and, especially, C. H. Wickens had done some work on the total mortality. As I had only occasional aid in the tabulations of mortality, I was grateful that Wickens had made a consolidation over the years 1911 to 1920 which resulted in a saving of about a quarter of the tabulations.

In 1971, I was happy to read a centennial oration in Wickens's honour; the story of his life from poor country boy to Commonwealth Statistician and leader in economic and demographic thought is remarkable. As I do not think even such addresses should neglect numerical data, I constructed a graph, which showed that *Demography*, the official annual, was published regularly in September with the data complete up to the end of the previous year in Wickens's time, which for technical reasons could not be improved. In immediate postwar times there was a lag of up to four years, and the introduction of sorting machines seemed to make little difference.

However, this is running too far ahead. I was awarded a Rockefeller Fellowship in Medicine to the London School of Hygiene and Tropical Medicine for 1947 but took it up in 1948, spending a year from August 1948 at the school. Professor Major Greenwood was still attending the school but was no longer interested in new recruits. Professor A. Bradford Hill was in charge of medical statistics but there were many others there who have also attained distinction: Dr. J. O. Irwin, Peter Armitage with whom I shared a room for a time, Dr. D. D. Reid, epidemiologist, and Dr. R. W. G. Doll, who was carrying out a survey on smoking and lung cancer with Bradford Hill. All of these are now well known to statisticians.

Among postgraduate students there were the biometrician, Dr. W. L. M. Perry, now Vice-Chancellor of the Open University, and J. H. F. Brotherston, now Professor of Community Medicine at Edinburgh. Under the Fellowship scheme there was no need to worry about degrees, but I attended as many meetings of the Royal Statistical Society as possible and also lectures and lecture courses at the School of Hygiene and at University College nearby.

Bradford Hill gave lectures to the Diploma of Public Health students. At University College, Dr. H. O. Hartley gave a good course on analysis of variance and E. S. Pearson on general statistical theory. In genetics, there was J. B. S. Haldane, who also gave a review lecture on Gaylord Simpson's book *Tempo and Mode in Evolution* to a packed theatre. I also attended courses from L. S. Penrose, who was very helpful when I suggested that Geissler's data were not sufficient to support the hypothesis that the sex ratio differed between human families.

As a result, I published in the *Annals of Eugenics* (later *Human Genetics*) a paper on sex-ratios in human sibships [8] which was attacked in a later number by C. Gini, who had written his Ph.D. thesis on the subject. A. W. F. Edwards took up the idea of Gini in an article in the *Annals* giving a large quotation from Gini, which caused me perhaps undue distress. Gini had considered it improper that anyone should criticize the work of German statisticians. However, the substantial correctness of my analysis is borne out by M. S. Teitelbaum on page 95 of the book *The Structure of Human Populations* [1]. He cites Edwards as writing "if genetic variability exists, it is of a very low order of magnitude." It is strange that the author who first came to this conclusion is cited in the references but not in the text! Incidentally, Armitage and I read carefully Fisher's discussion of the sex ratio in his *Statistical Methods* and agreed that it was a masterpiece of compression and good sense.

On the boat journey to London, I had written up some thoughts on the distribution of χ^2. I showed it to J. O. Irwin, who gave some good advice and help and forwarded it to *Biometrika* [4]. He obtained a modification of my partition and wrote it up as an accompanying article [2]. His advice as I was leaving London was that I should look for other interesting partitions of χ^2. Of course, I have followed his advice in a series of publications culminating in my book on χ^2 [24] and in my recent review-type article [27].

I was also working at this time on the theory of amoebic surveys and on blood counting. The first topic is of more general interest than its title suggests. It is also applicable to a study of the length of time before the first pregnancy of married women. Essentially, the problem consists of sampling from a mixture of binomial populations with parameter, p, ranging from 0 to 1 with a positive probability that $p=0$. A problem is to determine this unknown probability. In recent times, I discovered that Gini had also written on this topic.

In August 1949 I rejoined the temporary staff of the Sydney School of Public Health and Tropical Medicine, somewhat under a cloud, as I had enlisted support in England in an endeavour to improve my position in the school and to obtain clerical and computing assistance. I might mention that the relations between school and university were curious. Salaries at the school were paid by the Commonwealth Department of Health but some officers held university titles. Individuals, however, had no right to apply to the university for such titles. Some redress was made when on 2 March 1959 I became Associate Professor of Medical Statistics in the University of Sydney and in the school, but still not a member of the Faculty of Medicine. Before the closing date, 9 March 1959, I had applied for the newly created post of Professor of Mathematical Statistics in the University of Sydney and took up duties there on 17 June 1959.

7. My Research in the School of Public Health and Tropical Medicine, Sydney

Soon after my arrival back in Sydney, Dr. (now Sir) Kempson Maddox asked me to prepare a statistical survey of diabetes in the Australian population and in his diabetic clinic. Four articles [10], [13], [28] and [29] resulted, which are also referenced in my bibliography of vital statistics in Australia in the *Australian Journal of Statistics* [21]. In [28] it was shown that the recorded mortality rates from diabetes were high in Australia, and that under a commonly held genetic hypothesis the rates implied that over 30 per cent of genes were recessive. In [10], estimates of the prevalence of diabetes were made from ration cards; in [13], cancer of the pancreas and diabetes were shown to be positively associated and in [29] a follow-up study of the clinic patients was reported. The first of these articles reassured the editor of the *Medical Journal of Australia*, Dr. M. Archdall, that medical statistics had a place in medicine and thereafter he allowed me a practically free hand in submitting papers, indeed about 50 of them, to his journal.

I had had hopes of finding other clinicians who would cooperate in the manner of Sir Kempson Maddox but, after some disappointments, I decided that I would have to continue as sole author in almost all articles. In [31], I enlisted the aid of Mr. W. J. Willcocks, an official statistician, to describe the mortality data. Later, I cooperated with him in [33], a study of maternal mortality by age and parity of mother; New South Wales for many years had been the only large demographic unit in which these features could be traced; however, by 1951 interest had largely waned in maternal mortality, as it had fallen to less than a quarter of its value at the turn of the century and was about to fall further.

A number of the articles, [11], [15], [18] and [19] among others, dealt with the cancers. It was necessary to give age-specific and standardized rates for all the principal cancers; some special points arose; in [11] it was shown that cancer of the prostate was associated with marital status, an analogy with cancer of the cervix in the female. In [19], a clinical survey of no great extent showed that cancer of the lung was linked with smoking in Australia as in many other countries.

The most interesting find was that melanoma (black mole cancer) was associated with latitude in Australia. That it was associated with sunlight had been suggested by Dr. V. J. McGovern in 1952 but in [15] death rates from melanoma in the Australian population were given by state and hence by latitude. Clinicians and others had thought of sunlight as homogeneous but it was pointed out that the amount of ultraviolet light reaching ground level depended greatly on the altitude of the sun as there was greater filtration by the ozone layer in the upper atmosphere when the rays had to pass through it obliquely, whereas there was no such filtration with the visible light. The association was confirmed by a clinical survey [18].

It appeared that very little could be done to analyse separately the mortality rates for deaths from old age, uncertain causes, and cardiovascular, genito-urinary and nervous degenerative diseases. In my review paper on world mortality read to the Statistical Society of Australia in 1977 [26], I showed, indeed, that the total rates of deaths from this group of causes had stayed rather constant over the years. Publications by authors in Australia purporting to be able to measure the mortality from cardiovascular diseases have confirmed my impression that they cannot be so measured without extensive enquiries into certifying and coding practices.

In [6], [7], [16] and [17], I discussed tuberculosis in detail, giving especial emphasis to the importance of a generation or cohort analysis. Tuberculosis is a disease of long standing and the health of a man aged 60 years is decidedly dependent on his health over 40 years ago. Cohort analysis has not been well received by mortality analysts.

The era of my survey, say over the last hundred years, has been a happy one from a mortality point of view. Many of the infective diseases have disappeared and the importance of many others has greatly diminished. I have detailed the causes for these declines in [20] and [22]. It is difficult for some to see that many of these changes have been brought about by non-medical factors. One such factor is the ease of passage of the infective agent from one patient to another. This has been especially important in Australia for diseases like measles.

There was great interest in Australia over Dr. N. M. Gregg's discovery that many cases of congenital deafness had been due to maternal rubella in the first semester of pregnancy. An hypothesis was widely believed by which there had been a mutation of the virus to a strain which had this teratogenic effect. I was able in [9] to show from institutional data that similar events had occurred in the past; in particular, there had been excessive numbers of births of the deaf in Western Australia and South Australia in 1898 and 1899 and in New South Wales and Queensland in 1899. The epidemics in the Australian colonies could thus be explained as due to rubella fading out and then being reintroduced and causing infections in females carrying a fetus at the susceptible age; fade-out, of course, was due to the small size of the population. In papers [12], [14] and [30], a world-wide survey showed that epidemics of rubella deafness only occurred where fade-out was possible, showing the usefulness of stochastic ideas for looking at infections.

8. The Statistical Society of New South Wales and the Statistical Society of Australia

Soon after taking up my position at the School of Public Health, I organised a series of talks on medical statistics under the aegis of the Postgraduate Medical Foundation, 10 between 11 June 1947 and 13 August 1947 and 12 from 12 February 1948 to 27 April 1948, at an applied level with little mathematics. Miss Helen Turner was very helpful, giving five of the talks in the first set and two in the second set. We and others began to plan for a statistical society.

R. S. G. Rutherford had come out to Sydney University as Senior Lecturer by this time. A public meeting was held in 1947 under the chairmanship of Professor D. M. Myers and the Statistical Society of New South Wales was formed. Helen Turner and Stewart Rutherford gave papers at the first meeting and I gave a paper at the second meeting just before I left for England in July 1948. *The Bulletin of the Statistical Society of New South Wales* began to be published in March 1949. I joined the Council only on my return from England and was principally responsible for its publication until September 1958 when the last number appeared.

The *Bulletin* gave way in 1959 to the *Australian Journal of Statistics*, with myself as editor and H. S. Konijn and R. S. G. Rutherford as associate editors. The name was chosen in anticipation that there would soon be a Statistical Society of Australia. I remained editor of the *Australian Journal of Statistics* until my resignation from this position in 1971. Although I was keen to have the Australian journal, I was not confident that copy would flow in to us at a sufficient rate. I was encouraged, however, when Mr. D. W. Maitland was able to secure financial backing from commercial and other institutions. My fears about copy were justified when there was little material available for the No. 2 issue in 1964. I therefore submitted "A bibliography of vital statistics in Australia and New Zealand" [21] which filled pages 33 to 99. Several papers were then contributed by others which brought the *Journal* up to our required 148 pages for the year.

After that, the *Journal* has had no trouble with copy. It had been hoped that it would contain "applied" articles and the files show that this was so; but increasingly there were articles that appeared practical to those working in the field and "mathematical" to those outside it. Members of distant branches would then complain that they were losing members as a result of editorial policy. When this came up a second time after a year's break, I felt compelled to point out that my department gave more non-academic aid to the *Journal* in a month than the capitation fees of the whole far-distant branch produced for a year.

But by 1971 I had done my work, the *Journal* was well established and it was time to rest from patching up applied contributions and other chores and so I resigned. The only councillor who had supported me regularly became the new editor. However, I have never regretted my spell of editorship, 1959–1971; it would surprise "the general public" to know how much help and good advice an editor can obtain from near and far. With Australia's limited population, it was always necessary to avoid having local referees for local authors, and so it was an encouragement to receive willing help from referees abroad who were mere names to me at the time.

9. The Australian Mathematical Society

It will be remembered that the academic mathematicians in Australia in 1939 did not exceed two dozen in number and there were difficulties in travelling between states to attend the ANZAAS science congresses. After the war it was evident that there would be an expansion. Professor T. M. Cherry FRS and others were anxious to form a mathematical society in Australia. I was elected to the first council as an applied or practical (i.e. statistical) mathematician serving two years, 1956 and 1957. In 1959, Professor E. J. G. Pitman as President of the Australian Mathematical Society invited me to be General Secretary, in which capacity I served for four years.

By this time, papers were being received for the first number of the *Journal of the Australian Mathematical Society*. There had been doubts in some members' minds about whether there should be a journal; it was even suggested that we should run a, presumably experimental, journal for contributors under 30 or 35 years of age. However, the first editor, Professor T. G. Room FRS, left no doubt in referees' minds as to our standards; his advice was that "the standard of our Journal can be taken as equivalent to the *Annals of Mathematical Statistics*," and so on in other fields. The policy was completely successful and the journal is now well established.

I had three papers in the first volume. The first paper showed that the additive and multiplicative hypotheses in the tests in contingency tables sometimes led to the same test; the third showed that the absence of correlations between square summable functions ensuring independence in two dimensions could be generalized to several dimensions. The second devised a proof of the characterization of the normal distribution by the independence of linear forms in mutually independent random variables.

This was done without making any assumption on the existence of moments, in particular of the second moment, which had been assumed in all previous proofs. Cumulant theory was applied and the results followed with the aid of the theorems of Cramér and Marcinkiewicz.

10. Research and Teaching in the Department of Mathematical Statistics, University of Sydney

The Department was founded in 1959 as a result of the recommendation of the Murray Commission into Tertiary Education in Australia in 1957. I was, therefore, the Foundation Professor of Mathematical Statistics and only the sixth Professor of Mathematics in the first 107 years of the existence of the University. Dr. H. Mulhall, a fine mathematical scholar and possibly Australia's most experienced lecturer in statistics, joined my department from the older Department of Mathematics on his return from sabbatical leave; he continued to play an important role in the teaching and organization of the department.

It has always been my view that statisticians have used the methods of approximating to a discrete random variable, for example normal and χ^2 variables, in a faulty manner; a solution was offered in [5], in which the use of the uncorrected normal or χ^2 variables was shown to be equivalent to a Neyman–Pearson randomization.

Most of my mathematical work from 1959 onwards was on the application of orthogonal functions to statistical problems and included two joint papers, with M. A. Hamdan and with G. Eagleson. For a short time it could be said that, with those two authors and R. C. Griffiths, a "school" was in existence. The orthogonal theory has been summed up in my book, *The Chi-squared Distribution* [24] and in "Orthogonal models for contingency tables" [27]. In this second monograph, full use is made of the correlation generating function, which has very nice properties in the Meixner classes, namely the binomial, negative-binomial, Poisson, normal, gamma and inverse hyperbolic distributions.

This monograph and the note on the "Development of the notion of statistical dependence," reprinted from the *Mathematical Chronicle* (*New Zealand*) in [25] were with others part of a general survey which I had hoped to publish as a book. However, statistical dependence proved too large a topic to cover in one book.

I took up the invitation of the International Statistical Institute to write the *Bibliography of Statistical Bibliographies* [23]. By the time it was written, it was evident that a part-time secretary was necessary and that special attention would have to be paid to the sources of dates of the statisticians mentioned. I now have over 20,000 biographical cards of special design and can check entries from different sources. I have had thirteen addenda published to the book in the *International Statistical Review*. In recent years I have been able to extend the cards to general science as well.

The card file now contains all medallists, presidents and secretaries of the Royal Statistical Society, all Fellows or Members since their inceptions of the International Statistical Institute, the U.S. National Academy of Sciences and the French Académie des Sciences, all Fellows of the Royal Society of London since 1900 and so on. My author cards covering the greater part of the journal and book mathematical-statistical literature (about 70,000 cards together with a subject index) have now been taken over by the Division of Mathematics and Statistics, CSIRO, Canberra under the care of Dr. C. C. Heyde FAA, a member of my first fourth-year class in 1960.

11. Concluding Remarks

I have been fortunate in having many honours conferred on me during my career. Some of these are: Fellowship of the Australian Academy of Science in 1961 and its T. R. Lyle Medal for mathematics and physics in 1962; Life Membership of the Statistical Society of Australia in 1972 and its E. J. G. Pitman Medal in 1980; Presidency of the Statistical Society of New South Wales, 1952–3, of the Statistical Society of Australia in 1966–7 and Presidency of the Australian Mathematical Society in 1967–8; Honorary Fellowship of the Royal Statistical Society in 1975; Honorary Life Membership of the Australian Mathematical Society in 1981.

My degrees from the University of Sydney are MB, BS, BA, Ph.D., MD, D.Sc.

Now, after retirement from the Chair, my chief interest is in a survey of world mortality. A grant from the Australian Research Grants Committee is assisting me to complete a project which I have had in mind for many years, indeed, from the time when I was working on Australian mortality; further, in my sabbatical leaves of 1969 and 1973, I took advantage of the London libraries to prepare for this survey. The survey needs to bring together the official statistics, demography, some mathematical ideas on epidemiology and the field and laboratory epidemiology represented by Topley and Wilson's well-known work.

I am still extending my biographical index which I first began as an aid to the construction of the addenda to [23].

I look forward to an increased interest from statisticians in the use of test functions in multivariate analysis, general or non-normal, suggested by the additive theory; it appears to me that the models set up by the multiplicative theory as determined by the test functions are unnatural; on the other hand, the additive theory sometimes yields the parameters of non-centrality in a natural way. I am still working on the compatibility of joint distributions of subsets of a set of random variables, a difficult problem even when it is specialized to its combinatorial form.

It has been a great satisfaction that such interests in the history of science and of mortality have led me to take a heightened interest in the general history of mankind. A continued interest in some general problem of the world at large related to an individual's technical expertise can be its own reward.

Publications and References

[1] HARRISON, G. A. and BOYCE, A. J. (1972) (EDS) *The Structure of Human Populations*. Oxford University Press, New York.

[2] IRWIN, J. O. (1949) A note on the subdivision of χ^2 into components. *Biometrika* **36**, 130–134.

[3] KENDALL, M. G. (1943) *The Advanced Theory of Statistics*. Griffin, London.

[4] LANCASTER, H. O. (1949) The derivation and partition of χ^2 in certain discrete distributions. *Biometrika* **36**, 117–129 (Correction: **37**, 452).

[5] LANCASTER, H. O. (1949) The combination of probabilities arising from data in discrete distributions. *Biometrika* **36**, 370–382 (Correction: **37**, 452).

[6] LANCASTER, H. O. (1950) Tuberculosis mortality in Australia; 1908 to 1945. *Med. J. Aust.* **1**, 655–662.

[7] LANCASTER, H. O. (1950) Tuberculosis mortality of childhood in Australia. *Med. J. Aust.* **1**, 760–765.

[8] LANCASTER, H. O. (1950) The sex ratios in sibships with special reference to Geissler's data. *Ann. Eugen., London* **15**, 153–158.

[9] LANCASTER, H. O. (1951) Deafness as an epidemic disease in Australia: a note on census and institutional data. *Brit. Med. J.* **2**, 1429–1432.

[10] LANCASTER, H. O. (1951) Diabetic prevalence in New South Wales. *Med. J. Aust.* **1**, 117–119.

[11] LANCASTER, H. O. (1952) The mortality in Australia from cancers peculiar to the male. *Med. J. Aust.* **2**, 41–44.

[12] LANCASTER, H. O. (1954) Deafness due to rubella. *Med. J. Aust.* **2**, 323–324.

[13] LANCASTER, H. O. (1954) The mortality in Australia from cancer of the pancreas. *Med. J. Aust.* **1**, 596–597.

[14] LANCASTER, H. O. (1954) The epidemiology of deafness due to maternal rubella. *Acta Genet.* **5**, 12–24.

[15] LANCASTER, H. O. (1956) Some geographical aspects of the mortality from melanoma in Europeans. *Med. J. Aust.* **1**, 1082–1087.

[16] LANCASTER, H. O. (1957) Generation death-rates and tuberculosis. *Lancet* **273**, 391–392.

[17] LANCASTER, H. O. (1959) Generation life tables for Australia. *Austral. J. Statist.* **1**, 19–33.

[18] LANCASTER, H. O. and NELSON, J. (1957) Sunlight as a cause of melanoma: a clinical survey. *Med. J. Aust.* **1**, 452–456.

[19] LANCASTER, H. O. (1962) Cancer statistics in Australia. Part II. Respiratory system. *Med. J. Aust.* **1**, 1006–1011.

[20] LANCASTER, H. O. (1963) Vital statistics as human ecology. *Austral. J. Sci.* **25**, 445–453.

[21] LANCASTER, H. O. (1964) Bibliography of vital statistics in Australia and New Zealand. *Austral. J. Statist.* **6**, 33–99.

[22] LANCASTER, H. O. (1967) The causes of the declines in the death rates in Australia. *Med. J. Aust.* **2**, 937–941.

[23] LANCASTER. H. O. (1968) *Bibliography of Statistical Bibliographies*. Oliver and Boyd, Edinburgh.

[24] LANCASTER, H. O. (1969) *The Chi-squared Distribution*. Wiley, New York.

[25] LANCASTER, H. O. (1972) Development of the notion of statistical dependence. *Math. Chronicle* **2**, 1–16.

[26] LANCASTER, H. O. (1978) World Mortality Survey. Address to the Statistical Society of Australia, 12 July 1977. *Austral. J. Statist.* **20**, 1–42.

[27] LANCASTER, H. O. (1980) Orthogonal models for contingency tables. *Developments in Statistics* **3**, 99–157.

[28] LANCASTER, H. O. and MADDOX, J. K. (1950) Diabetic mortality in Australia. *Med. J. Aust.* **1**, 317–325.

[29] LANCASTER, H. O. and MADDOX, J. K. (1958) Diabetic mortality in Australia. *Aust. Ann. Med.* **7**, 144–150.

[30] LANCASTER, H. O. and PICKERING, H. (1952) The incidence of births of the deaf in New Zealand. *N.Z. Med. J.* **51**, 184–189.

[31] LANCASTER, H. O. and WILLCOCKS, W. J. (1950) Mortality in Australia: population and mortality data. *Med. J. Aust.* **1**, 613–619.

[32] NITSCHKE, R. A. (1980) *The History and Family Tree of Johann Wilhelm Rohrlach and his Wife Anna Dorothea née Galpach and their Descendants, 1792–1980*. Lutheran Publishing House, Adelaide.

[33] WILLCOCKS, W. J. and LANCASTER, H. O. (1951) Maternal mortality in New South Wales with special reference to age and parity. *J. Obstet. Gynaec. Brit. Emp.* **58**, 945–960.

Index

A

B

C

F

G

H

K

L

M

N

O

P

Q

R

S

T

U

V

W

Y

Z